Charles Cutter

Guide to Niagara Falls

Charles Cutter

Guide to Niagara Falls

ISBN/EAN: 9783742810205

Printed in Europe, USA, Canada, Australia, Japan

Cover: Foto ©berggeist007 / pixelio.de

More available books at **www.hansebooks.com**

NIAGARA FALLS,

AND ADJACENT POINTS OF INTEREST.

CUTTER'S GUIDE to THE HOT SPRINGS OF ARKANSAS. 3D Edition, 114,000 issued; MT CLEMENS, MICH. 15 Editions, 68,000 issued; THE MINERAL WELLS OF TEXAS, THE EUREKA SPRINGS OF ARKANSAS; THE THERMAL WELLS, CITY OF WACO, TEXAS.

Fully Illustrated.

PRINTED BY HEALTH AND PLEASURE RESORTS

Niagara Falls.— General View from New Suspension Bridge.

NIAGARA FALLS.

THE FALLS.

OVER five hundred thousand visitors have the pleasure of viewing the Falls of Niagara every year.

The many millions who annually pass by, either via Buffalo or Suspension Bridge, without stopping over at Niagara Falls, certainly cannot fully realize that they are losing the opportunity, possibly of a lifetime, of beholding one of the grandest and most beautiful works of nature.

Standing upon the brink of the precipice at Prospect Point, and taking the first good view of the Falls, every one is impressed with their majestic beauty and enormous power, and when they contemplate the immensity of the volume of water, the great depth of the chasm into which it makes its mad though graceful plunge, a mingled feeling of awe and admiration takes possession of the beholder.

It is not our mission to attempt a flowery or classical description of Niagara Falls, but rather to guide the reader to the many points of observation from which they can be seen to the best advantage. By presenting a larger number of beautiful illustrations than were ever before issued in any one publication, we seek to induce as many as possible to come to Niagara, and enjoy for themselves the pleasure of seeing the greatest cataract known to the civilized world.

The finest writers in the English language—Dickens, Trollope, Thackeray and a host of others—have expended their best efforts at word-painting in attempting to set forth the beauties of Niagara. But words seem weak and powerless, and the great cataract baffles description. We can only hope, therefore, to make our descriptions acceptable to the reader by supplementing them with profuse illustrations. These are half-tone reproductions from the best photographs obtainable from leading photographers, or made by our own artist.

But there are a few facts and figures to which we gladly call the attention of our readers, that they may more clearly understand the power and magnitude of Niagara. It is by comparison only that we can fully comprehend them. It has been estimated that the average amount of water flowing over the Falls is 3,600,000 cubic feet every twenty-four hours, which is said to represent a force equal to the same amount of power produced from two hundred thousand tons of coal; the daily output of all the coal mines in the world.

When we consider that competent engineers estimate the theoretical power of these Falls to be from five million to six million horse power, and that all the water mills of the United States have only developed an amount equal to about one-fifth of this immense force, we are again led far beyond ordinary

Birdseye View of Goat Island and American Rapids.

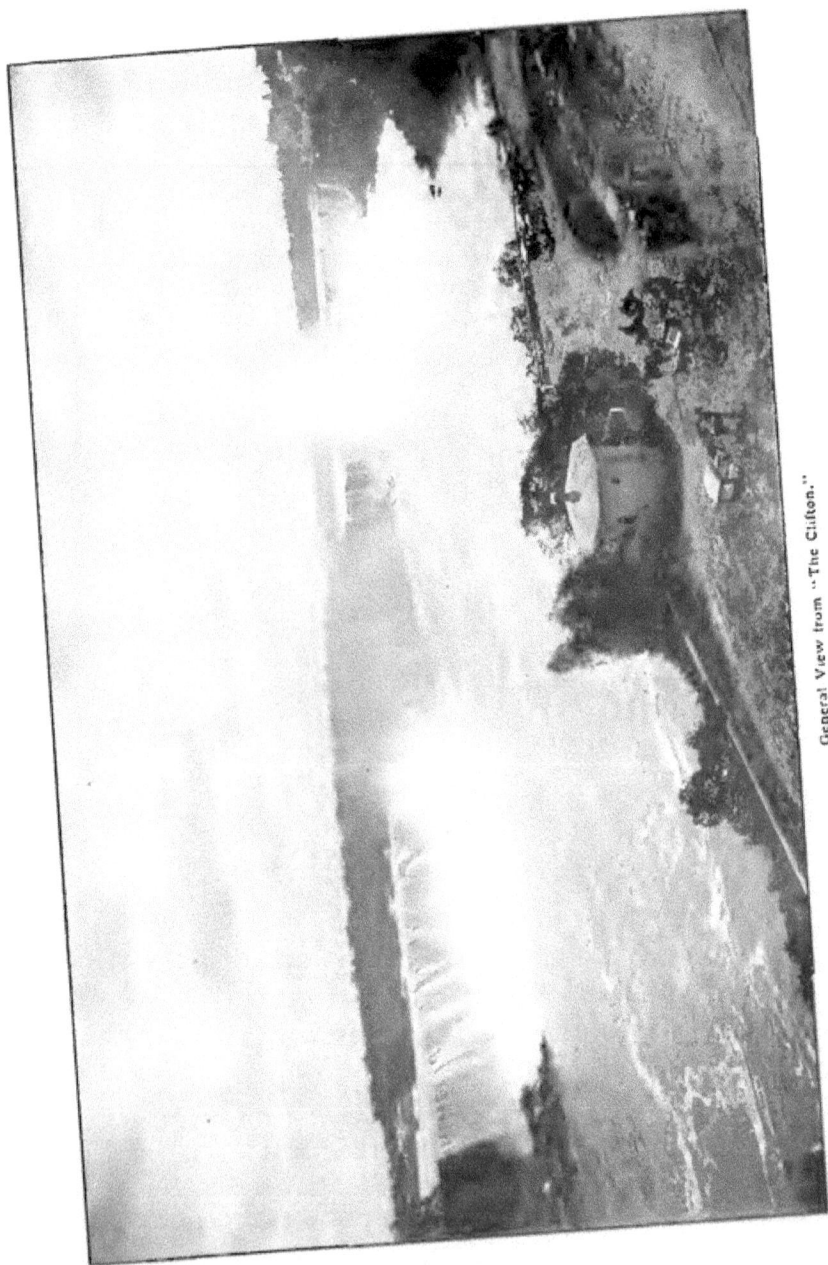

General View from "The Clifton."

American Falls, from Below.

comprehension, as to the great power of these grand and beautiful Falls.

The American Falls are about 1000 feet wide, including the Bridal Veil or Central Fall, and the height is 164 feet, the rapids above these Falls have a fall of forty feet within one half a mile. These falls are more permanent than are the Horseshoe Falls, the recession for the last half century being hardly perceptible, probably owing to the fact that the volume of water is not sufficient to remove the large boulders upon which it descends, and they in turn protect the under stratum of soft shale from washing away.

The Horseshoe Falls, frequently called the Canadian, have a contour of about 2,750 feet (the survey of 1886 showed them to be 2,600,000) and are 158 feet high; the Canadian rapids have a fall of fifty-five feet in three-quarters of a mile before reaching the crest of the Falls. The volume of water passing over this Fall is estimated to be from three-quarters to four-fifths greater than that over the American, and the depth of the water at the crest to be from two to twenty feet (one to fourteen we think to be nearer correct). The recession of this Fall is very rapid, and those who have known them from twenty-five to fifty years notice a very material change. The name was first given to this Fall on account of its form resembling that of a horseshoe, but it is so no longer, as it now more closely resembles the letter V; and the recession which has caused this change of form is slowly destroying its beauty, as the greater volume of water is attracted to the center and is thus gradually being drained from the sides.

Geologists say that this recession, if continued at the present rate, will eventually drain the American Falls and leave what is now the American rapids a bare, rocky strip between Goat Island and Prospect Park. But as this will probably not take place until after the recession has caused the Falls to pass Goat Island, several thousand years from now, we trust that none of our friends will defer their visit to Niagara Falls on that account.

We desire to call the attention of every visitor to the "air explosions," which occur at intervals of from five to twenty-five seconds, those showing the greatest force and most readily perceived being from the center of the cataract. This phenomenon is of recent discovery, and it is to this that the greatest cause for the recession is now attributed. Further mention of the subject will be found elsewhere in this work.

The Gorge, Victoria Park, and "The Clifton," from New Suspension Bridge.

HOW NIAGARA CAN BEST BE SEEN.

OO many come to this resort expecting to see all that to their minds is worth seeing, in a very few hours, and then hasten to the station to catch the next train. These people are more wise than those who are satisfied with a hurried glimpse of the Falls from trains which pass in sight, but at a distance, but do not stop off even for an hour; and even these are wiser than those who pass from east to west, or *vice versa*, and do not select a route that will take them within sight of this great cataract.

A mere sight of the Falls from the car window is far better than not to see them at all; but a few hours spent at the Falls is a thousand per cent better yet; though the wise ones who spend several days here, have the pleasure of seeing them in all their localities and from the best points of observation, and are thus able to fully appreciate them.

To intelligently decide how best to see Niagara, it is necessary to determine how long a time can be spent at sightseeing before planning what to see first. We will therefore give a description of the routes and bytrips for the benefit of those who conclude to remain several days, confident that many of our readers who planned only for a few hours' stay will continue their visit longer.

After being comfortably located at your hotel, take a walk or ride to Prospect Park, and a good view of the Cataracts from Hennepin View and Prospect Point; at each of these places you can spend twenty minutes of an hour enjoying a feast for the eyes well worth traveling around the world to behold.

If your arrival is in the morning, after taking in these views you will have time enough for the sights in and about Prospect Park to occupy your mind until luncheon or early dinner; if it is afternoon, and these sights have been enjoyed, and after a short ramble around the Park, your mind and body can well afford to rest for the day; but we would not

be surprised if you desired to see the Falls again, possibly by twilight or moonlight.

The points of observation mentioned are usually the first to which the visitor is taken when riding, but as the time required for a satisfactory view from these places, which are considered the best on the American side, from which the Falls, the Gorge and the surrounding scenery can be admired, it should be as near an hour as circumstances will admit, and to keep a carriage in waiting, at an expense of one dollar per hour, is a matter which some may be disposed to consider.

The next trip will be over the bridges to the Islands. As this route is a little more than two miles from Prospect Park, and a number of stops should be made in order to do it well, at least two hours' time should be devoted to it; our readers can best determine for themselves if they prefer to ride or walk. The Park carriages make this trip, giving the privilege to stop over at any point on this route and to resume seats in another of the carriages. The first is Bath Island, reached by a bridge over the rapids, thence by another bridge to Goat Island, so named because at an early day a number of these animals were pastured here. The popular drive around this Island is to the right, and the following are the points of interest to be visited, in the order named, and as much time can be devoted to each as the individual visitor may deem necessary:—

Luna Island is reached after a short drive or walk, thence by a path leading down by the sides of the bank to the bridge, or a stairway from Steadman's Point leading to the same, (better take the path and return by the stairway). At Luna Island fifteen minutes time can be very pleasantly spent.

On returning to the cliff by the stairway, you are at Steadman's Bluff, from which several grand views are to be

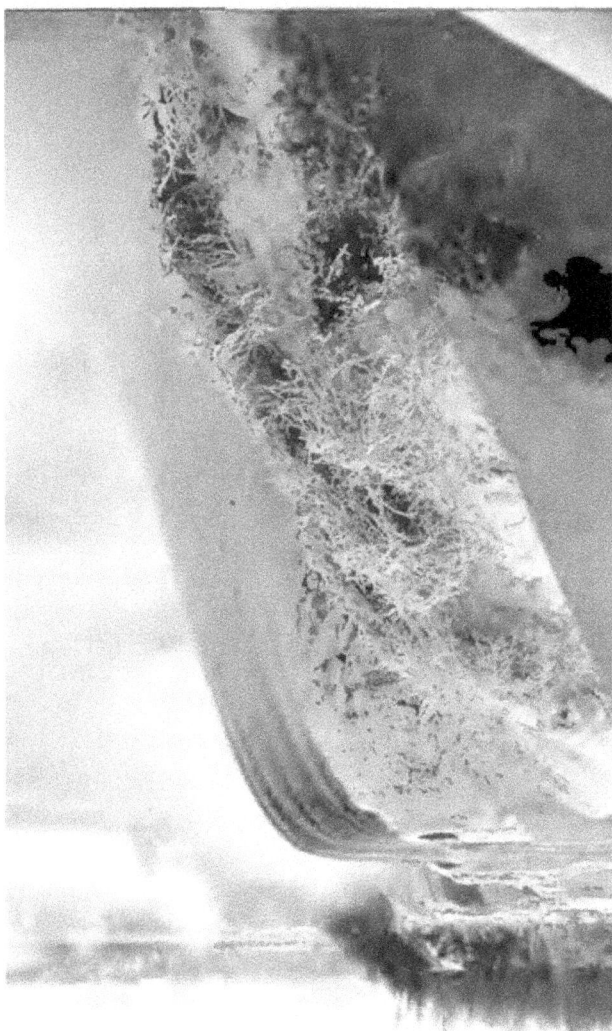

Prospect Park and Island Bridges.

had. One or two hundred feet further on, following the line of Steadman's cliff, brings you to the office of the "Cave of the Winds," and twenty minutes or half an hour can be spent here in taking the romantic trip through and around the cave. Do not fail to visit it if you can spare the time. and the dollar expense necessary. The drivers of hacks and carriages may not encourage this, as they receive no commission from the receipts, but do not heed them, even though they advise otherwise, as the trip is an adventure and experience well worth the time and expense.

Continuing along the road which skirts the cliff a few hundred feet, we reach Porter's Bluff, where the first near view of the Horseshoe Falls is to be had on the Island by the route taken. From here a stairway leads down to the path to Terrapin Point, where many years ago was built a stone tower, which after long and constant use was torn down because it was considered dangerous. At this point it will be well for the visitor to observe the little signboards of the commissioners, which read, "Do not venture in dangerous places." But this need not intimidate the visitor so much as to prevent the enjoyment of viewing the grand sights before him, as this is the nearest accessible point to the center of the Horseshoe Falls, and the scenery is grand beyond description.

Returning to Porter's Bluff, the road is continued along the edge of the rapids, four or five hundred yards, to the entrance to the Three Sister Islands, which are reached by the three bridges, photos. of which we have reproduced in the group on the opposite page. On the Sister Islands a half hour can be very pleasantly spent viewing the little cascades between the islands and the great Canadian rapids, of which a grand sight is to be had from the third island of this cluster.

Returning across these bridges to the road on Goat Island, it is followed around the upper end, which it circles to the point where the waters divide; part going to the American and the greater quantity to the Horseshoe Falls; from here we follow the former towards the rapids, and all

the way to the place of entrance on the island, noticing the pretty views of the Park and the city, and watching the rough and tempestuous waters, which but a few moments before we witnessed calmly and peacefully parting from their sweethearts with a kiss; now in their mad, wild fury they are hastening to be the first to make the desperate leap of the cataract and to meet their loved ones in the deep chasm of the gorge below the Falls.

With the return to Prospect Park, or the city, ends route number one.

For route number two, we will take a trip over the new Suspension Bridge to Ontario. The fare over this bridge is 15 cents for the round trip or 10 cents for one way. From the bridge one of the very best general views of the two Falls, Goat Island, the Parks and adjacent scenery is to be had. The floor of this bridge is one hundred and ninety feet above the water, and the sensation in crossing is one of awe, which is speedily dispelled in admiration of the surrounding scenery.

After crossing the bridge and passing a short distance to the left, Victoria Niagara Falls Park is reached. From its borders on the cliff a magnificent view of the Falls is to be seen: probably the best from any point, as it is immediately opposite the American Falls, and affords the best general sight of the Horseshoe Falls as well. The place from which this grand view is to be had is named Inspiration Point. If this trip is taken on foot, numerous places can be found where seats and shelter are provided, and where ample time can be spent in feasting upon the grandeur of the scenery.

If riding, the scenes can be enjoyed while the ride continues through the Park to the Dufferin Islands, a romantic locality, as the names of its "Lover's Retreat," "Rambler's Rest," "Lover's Walk," etc., fully attest. For the walk or ride through Dufferin Islands a fee of 10 cents is charged for pedestrians, 50 cents for conveyances drawn by two horses, and 25 cents for those drawn by one horse.

One half hour or more can be enjoyed on these islands.

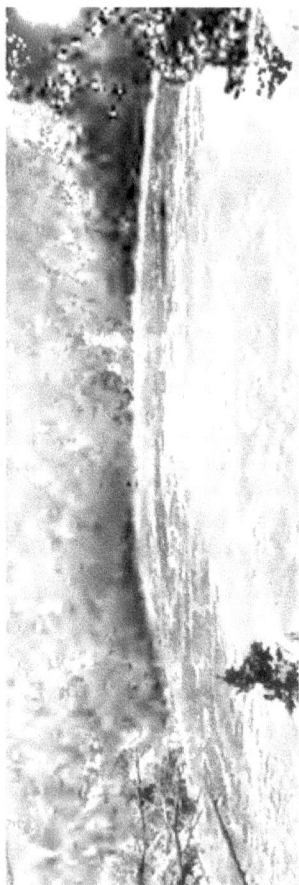

At the end of Lover's Retreat is to be seen the grandest view of the Canadian Rapids, which are one mile wide, and nearly the same distance in length. Many of the big waves are to be seen dashing their spray twenty feet high, reminding one of the seashore in rough weather, and making a panorama of which one seldom tires.

A short distance beyond these islands is the Burning Spring, well worthy of a visit, and for which a fee of 50 cents is charged; it is a strong sulphur spring from which a large quantity of gas escapes. This is ignited by an attendant, first as it flows through a long gas pipe, which when lighted makes a brilliant blaze; and afterwards as it bubbles up through the water in due spring, setting fire, as it were, to the water. The spring is situated in a darkened room, making a very pretty sight as the flame illumines the darkness.

From here those in conveyances can return by the same route or over a Canadian country road, passing Loretto Convent, through the old village of Drummondsville, a pretty suburb of Niagara, to the Lundy's Lane battlefield. Here a tower has been erected from which, for a fee of 25 cents, an excellent view of the surrounding country can be seen: thence returning by the old village of Clifton, new Niagara Falls South (Ontario), then over the new Suspension Bridge to the hotel. Thus ends route number two.

THE TRIP DOWN THE GORGE.

THE deep and indescribably wild gorge, which in the centuries past has been worn into the rocky formation of the earth's surface, beginning at the Lewiston Mountain or the Niagara escarpment, and winding backward to the Falls, is one of the most interesting places in the world. The early history of this gorge is filled with blood-curdling incidents, thrilling struggles for supremacy between the Indians and the settlers, and the roar of musketry of the battles between French, Indians, English and Americans as they each struggled to dispossess the other of this valued territory. Many places along the banks at the verge of the gorge show evi-

dences of having once been the scene of battle, and even now traces of earthworks of improvised fortresses are to be found. The Niagara gorge is seven miles in length from the Falls to Lewiston, and its depth above the water level increases from 164 feet at the Falls, to 300 feet five miles below. Nature has prepared many wonderfully beautiful and picturesque scenes for the visitor; scenes that can be viewed many times before the eye wearies of the repetition.

Directly below the waterfall, the water of the Niagara river is variously estimated to be 100 to 250 feet in depth in the path in the main channel, and for nearly two miles, about one-quarter of a mile in width. Here it moves along slowly, and is navigated by the steamer "Maid of the Mist."

The cars of the Niagara Falls and Lewiston Railway, known as the "Gorge Route," which start from Prospect Park at Falls street every fifteen minutes during the season, enter the gorge about half a mile above the railroad bridges and the head of the Whirlpool Rapids, and as the cars wind their way downward along the serpentine tracks of the incline, the passenger sees below him the great gorge; to the south a most beautiful view of the Falls in the distance, and to the north the mammoth cantilever bridge of the Michigan Central and the new steel arch bridge of the Grand Trunk railway. As the bottom of the incline is reached, the gorge suddenly narrows, and the immense volume of water, forced into the narrow defile, bursts into a fury and roars wildly as it dashes the waves high in the air and at a terrible speed rushes toward the whirlpool. The transformation seems almost incredible. Passing "Rapids View," under the bridges, then the Whirlpool Rapids, and "Observation Rock," the tracks wind along beside the mad waters to the Whirlpool, where a stop is made. The peculiar phenomenon of the waters caused by the sudden changing of the currents and the mad rushing stream from above, combined with the beautiful picture presented by the verdure-covered rocks of the ravine, makes a stop well worth the time spent.

From the Whirlpool the river surges away to the north again; beginning here what is called the "Devil's Hole Rapids," and tumbling and leaping through the narrow ravine, each

The New Single Arch Steel Bridge, Grand Trunk Railway.

Horseshoe Fall and Islands, from Canada.

(21)

American Rapids.

wave jostles the next over the big rocks that make the stream at most perilous one, and with a current almost equal to that of the Upper Rapids. The picture is an inspiring one, and not to be forgotten. Three hundred feet above, are to be seen the grotesque faces of the rock, seeming to threaten the adventurous ones below; beside the tracks the waters lashed into foam, and beyond, the Canadian shore, less steep, and covered with nearly every known species of tree and flower in its native state. Below the Whirlpool is Brinker's Park, a shady, moss-grown resting place, named after Capt. Brinker, the founder and president of the road. A little beyond this park and just opposite the historical "Foster's Flats," on the Canadian shore, is Giant Rock, a monster boulder that stands aloof from the mountain, outside of and overhanging the railroad. The "Sentinel," another large boulder, projects far above the rushing waters of the rapids nearby. The "Devil's Hole" beyond this point is the most interesting and historical of all stopping places. From the car the ravine of the Bloody Run may be seen, and 300 feet above, the "Devil's Pulpit," from which ledge 200 British soldiers were hurled over the bluff by the Seneca Indians. Winding along the rugged shore the tracks pass many interesting ravines, over streams, the waters of which are whitened by the minerals from the rocky caves where they flow, and finally under the old earthworks of Fort Gray, which was occupied by the American army during the war of 1812, and opposite which is seen the towering "Queenston Heights," where stands the monument to Gen. Brock, killed in the Battle of Queenston in 1812. From mountain to mountain are stretched the few remaining strands of the Lewiston suspension bridge, wrecked many years ago by severe winds, and which was once the crossing place of travelers between New York and the west. The Gorge line here emerges from the deep cañon, upon the broad plateau through which the river sweeps majestically into Lake Ontario seven miles north, and the traveler finds himself in sleepy old Lewiston, where the fisherman, the lover of the historical and the student of nature may all be satisfied alike. This trip may be taken in 30 minutes, returning in the same time. The better way is to secure stop-over checks (which are issued without extra cost) for Rapids

View, Buttery Whirlpool Rapids, Brinker's Park and Devil's Hole, in order to study the beauty and grandeur of these places. This round trip costs sixty cents. We would advise the visitor to follow the Niagara to the lake for a visit to old Fort Niagara, Youngstown, and beautiful Niagara-on-the-lake. The Lewiston & Youngstown Frontier electric line makes direct connections with Gorge cars, carrying passengers to Youngstown. Round trip thirty-five cents. Ferry to Niagara on the Lake fifteen cents each way. Steamers of the Niagara Navigation Co. ply between these points and Toronto, electric cars connecting.

A TRIP TO OLD FORT NIAGARA.

ONE of the most interesting short journeys to be taken out of Niagara Falls, is via the Lewiston and Youngstown Frontier Railway, from the river front at Lewiston, where connections are made with the Gorge cars, the New York Central trains, the steamers of the Niagara Navigation Company and the Queenston ferry. The "Old Fort"

Wreck of the Old Suspension Bridge.

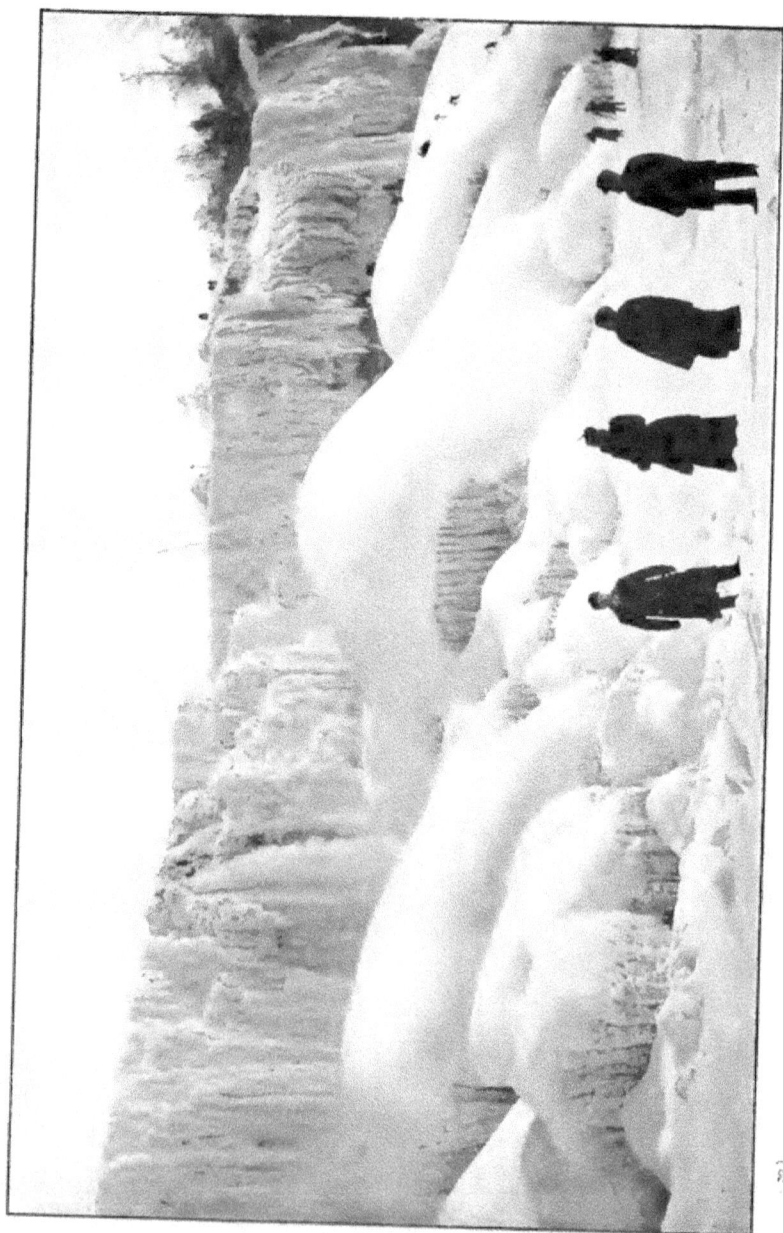

The American Falls in Winter.

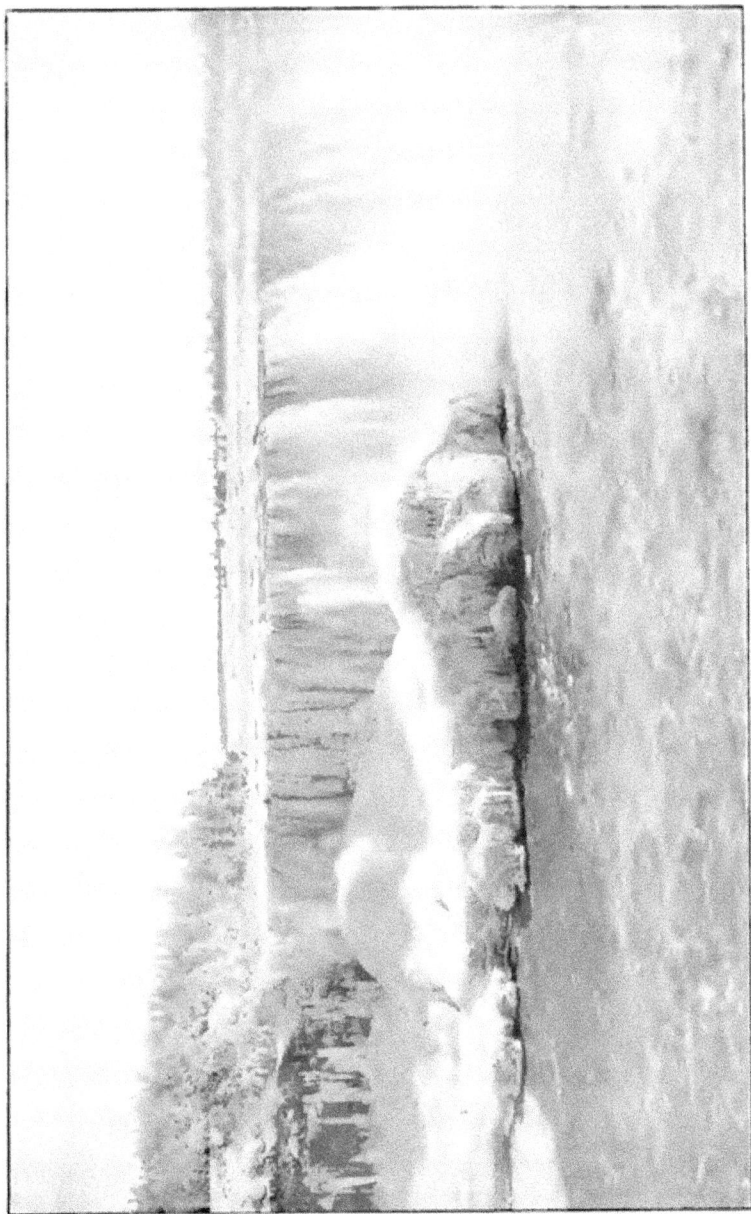

The Horseshoe Falls: Early Winter.

The American Falls from the Canadian Shore.

Old Fort Niagara.

The Old French Castle.

"Ronte," as it is known, winds up the hill into Main street of historic old Lewiston, and the visitor is in the heart of what was at one time the greatest port of entry on this frontier, and which, from the earliest history known, has been the most important portage of the Niagara, being on the direct line between New York and Detroit. Far up on the bluff to the north, stands the old Barton homestead, the first residence on which was erected by Major Benj. Barton, whose appointment in the American army was made by President Madison, and who for many years associated with Gen. Porter and Judge Porter in the ownership of much of the property along the frontier. An interesting feature to-day is the old elm tree which guards the entrance to the gate at the homestead, and about which is woven a pretty romance. This tree was once a riding whip in the hands of the charming young wife of Major Barton, on the day when she rode on her favorite saddle beast from Geneva to join her husband at her new home

goon of the old castle, and Porter's History of Fort Niagara gives a detailed account of the disappearance of Morgan and the consequent political strife in New York state.

Opposite the fort is to be seen fort Mississauga and Niagara-on-the-Lake, a popular Canadian summer resort, and near the fort is said to be the best fishing grounds on the Niagara frontier. On a clear day the scenery surpasses that of many western valleys, and this adds intensity to the pleasures of the journey. At the terminus of the road is a large grove of 30 acres which is used as a picnic ground.

THE WHIRLPOOL RAPIDS.

BEGINNING at a point just below the great railroad bridges, the Niagara river which, below the Falls and to this point, moves along, in a sleepy way, is compressed into a narrow defile and begins its descent to the lake twelve miles below, along a declining and rocky bed which lashes the water into an indescribable fury for many miles. The distance from the head of the Rapids to the Whirlpool is one of the most turbulent and treacherous waterways in the world and but two floating craft ever dared to attempt the journey down the incline. The feat was successfully accomplished by the old steamer "Maid of the Mist," several years ago, and a life boat built by Mr. C. A. Perry, of Suspension Bridge, who made the trip successfully. Two human beings have attempted to stem this awful tide, and one, a local adventurer escaped alive, although insensible from the buffeting of the waves and only prevented from sinking by a liberal quantity of cork life belts. The other, Captain Matthew Webb, the once famous English swimmer, attempted the trip on the afternoon of July 24, 1883, and met death in the waves, it being generally believed that life was crushed out of the luckless fellow in the monster waves that rise to an estimated height of thirty feet, at brief intervals, just opposite the Buttery elevators on the American side and the incline railway on the Canadian side. The

---

early in 1806. The madame, half in jest, planted the whip on the hillside and it stands to-day, gnarled and knotted, having escaped the destruction which followed the invasion of the British and Indians in the war of 1812. On, through the scenes of the historic struggles between the French, British, Indians and Americans, and past the old Frontier House, which since its erection in 1824-5, has been until recent years the most important hostelry in this region. Many brilliant assemblages have been seen in the once richly decorated ball room, and men who have made the history of the world have been its guests. This was the stage coach headquarters for Western New York, and in the Masonic Hall, on the upper floor, it is said William Morgan was tried for infidelity to the order.

At the turn of the railway may be seen, a short distance further east, the ancient Lewiston Academy, long since abandoned, but which was the seat of learning in Western New York in the early part of the century. The trip continues through the finest of farm lands and endless orchards laden with the choicest of fruit, and into the town of Youngstown, where for one hundred years there was no cessation of blood-curdling events. After passing the steamer landing and the El Dorado Hotel, the line enters the confines of old Fort Niagara. Here in 1678, LaSalle, LaMotte, Ft. Hennepin and fourteen other French adventurers landed from Lake Frontenac (Ontario), and in the midst of a Seneca Indian village established a trading post, which later on was converted into a block house. This being the key to the possession of the lake region, the French made several unsuccessful attempts to gain the consent of the Indians to establish fortifications, and in 1720 erected, partly through deception, a block house at Lewiston, five years later by permission, establishing what is now known as the old "French Castle," the first stone building at Fort Niagara.

During the war of the Rebellion many of the most destructive of the British expeditions were sent out from Niagara. It is said that William Morgan, the offending member of the Masonic fraternity, was last seen alive in the dun-

from the Rapids View platform just below the bridges, or from the Canadian shore, which is reached by an inclined railway.

THE DEVIL'S HOLE

NO more interesting or historic place is to be found than this weird and enchanting spot about three miles below the Falls. Here a grisled old rock projects out over the gorge below, keeping guard on the rushing waters of the Devil's Hole Rapids that roar and wind in and out along the picturesque valley known as the gorge of Niagara. This monster rock is 300 feet above the water, and the top is as

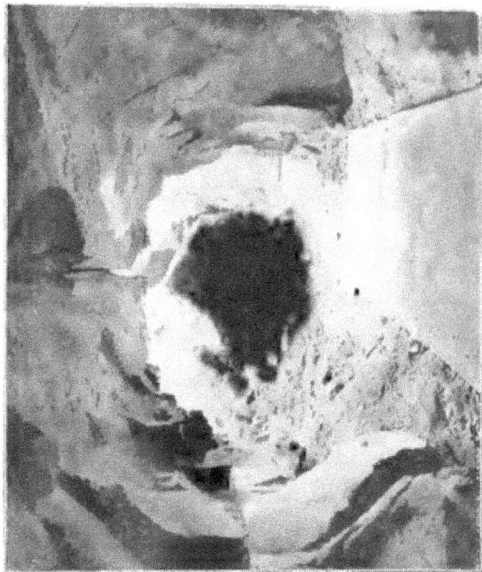

"Devil's Hole."

Whirlpool Rapids.

water of the Rapids is said to be very deep and travels at the rate of thirty miles an hour, the terrible force swinging the main current backward and forward from side to side like a drunken thing, careening one wave against the next, until the spectacle becomes at once grand and awe-inspiring.

The mad rushing of the waters churns and dashes the waves first heavenward in snow white spray, then mercilessly against the rocks that jut out into the boiling stream, until, pounded into a living mass of foam, it empties into the Whirlpool three-quarters of a mile below the bridges.

To get the most comprehensive view of the Rapids the visitor should spend several minutes, the longer the better, on the Battery observation rock, which is on a natural level with the water just at the foot of the Battery elevators. No fee is charged for occupying the rock, but to travel up or down the elevators a distance of 300 feet, a fee of 50 cents is charged for the round trip. The Rapids may be viewed

off the precipice or murdered them with tomahawk or knife, until the little stream that fell in beautiful cascades down into the ravine beside the towering rock, ran red with the blood of the victims. To this day the big flat rock is known as the Devil's Pulpit and the stream as Bloody Run. Historians differ in stating the number of British killed in that massacre, but it is generally agreed that more than 200 men were in the detachment. John Stealman, commander of the wagon train, escaped on a fleet-footed horse, a drummer boy named Matthews fell into the forked top of a tree in the gorge and escaped, and a wounded wagon man escaped by hiding in the bushes. A detachment of soldiers in camp at Lewiston marched to the scene and met the same reception at the hands of the red men, only eight escaping to tell the tale.

A winding stairway now leads into the ravine, a fee of 25 cents being charged for admission. The ravine is one of the most refreshing spots to be found, and the narrow walk along the face of the rocks takes the visitor to the Devil's Hole, now a cleft in the rock about 30 feet deep, and at the further end of which is a spring from whence the sweetest of cold water flows the year round. It is necessary to stoop low to enter the cave, but several persons may stand erect within the main chamber. In front of the opening stands a large boulder called Ambush Rock, and which at one time covered completely the mouth of this retreat, making an impregnable fortress. It is stated by the older settlers that before the engineers of the New York Central railroad began blasting in the vicinity, the Devil's Hole cave

(26)

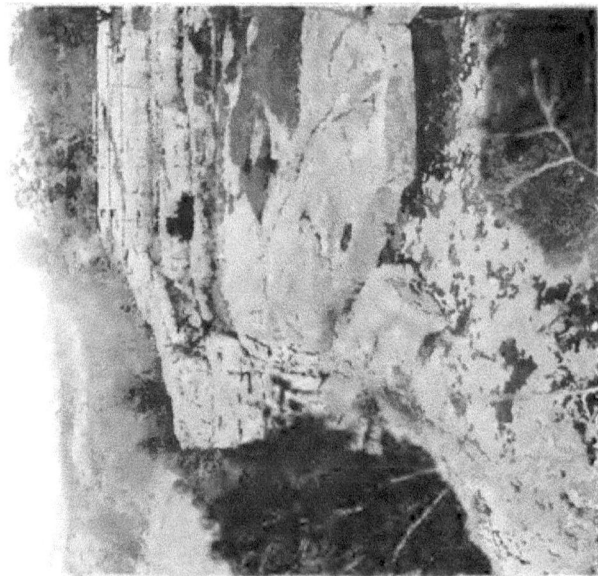

"The Devil's Pulpit."

extended three quarters of a mile into the rock and had often been explored. The interior of the cave appeared to have been cut out by rough instruments, and it is believed was formed by some pre-historic race. A winding walk leads to the old Council Rock at the head of the long stairway, and around which it is said the friendly chiefs of the Five Nations for many years held their councils or "powwow's" over the affairs of state. A pathway almost hidden by moss takes the adventurous to the lost channel of Bloody Run, where the little stream disappears under the rocks, but which was once a beautiful cascade. In this ravine and the bed of the stream, relics of the battles of the eighteenth century are often found. Stairways lead to the rapids and tracks of the Gorge railway below.

INTO THE HEART OF NIAGARA.

ALTHOUGH many thousands of visitors to Niagara Falls view the great cataracts from above, and from many points, yet it is truly said that no one has seen Niagara in all its glory, or can accurately measure or understand the power of the great waterfall, until he has taken a trip on the little steamer "Maid of the Mist."

To reach the steamer from the American side there is a long incline with free stairway for foot passengers, and comfortable cars upon which passengers are carried for ten cents down and back. The landing on the Canadian side is reached in a similar way. On board the steamer, the passenger is furnished with a comfortable rubber suit, and, encased in waterproof from head to foot, he is stationed on the upper deck to drink in the beauties of the scenes about to transpire.

The feeling that possesses the passenger on board the little steamer as she enters the great cauldron at the foot of the Horseshoe Falls is apt to be one of timidity, and he instinctively draws back from the rail of the boat and closes his eyes as the little boat faces the torrents of spray and rides over the seething water into the very heart of Niagara. But there is no

danger; and when in a few seconds more the steamer floats backward upon the foam whitened waves, and the passenger realizes that before him is one of the grandest pictures ever painted by Nature and one which defies the brush of the most skilful artist to reproduce, he experiences a sense of pleasure when for the second time the bow of the steamer is turned toward the falls and enters again.

Above, nearly 200 feet, the river empties her green water over the brink of the precipice, and as it breaks into foam in its terrific descent, it gives forth a roar, the power of which is indescribable. Into this storm center the little boat ploughs her way, and to the very verge of the roaring, seething mass of water that falls with resistless force into the pit below. The spray clouds dash about from side to side, and the compression of air often sends the spray high above the great falls, making the picture the more beautiful. Below, the water is churned into a river of "black and green and white, a boiling

Steamer "Maid of the Mist."

(27)

The Cave of the Winds in Winter.

Winter View, American Falls from Goat Island.

(32)

The Ice Mountain.

The Ice Bridge.

stream of molten malachite," and as it escapes from the cauldron whence it falls, it boils and seethes like a mad thing as it tumbles about and rolls away into the stream beyond. Then the pilot turns his craft away, and the passenger, at first so timid, leaves behind him the grandest of scenes, and as he watches the picture dissolve in the distance, resolves to repeat the experience as often as opportunity offers.

The steamer is a staunch little craft and rides the turbulent water like a feather. As she passes over the American Falls the great white clouds of mist dash over the boat, and sometimes seem to envelop her completely. The bright sun peeping through the mist forms little rainbows, and each passenger sees, starting out from beside his very feet, a tiny colorbow, dancing away in the clouds and returning again until one is encircled as in a halo of glory. As the boat leaves the first falls, the bow becomes larger and larger, until it is swallowed up in distance and one sees above him the monster shafts of roaring water tumbling one over the other in their rollicking way.

The steamers are under careful management, and are annually licensed and inspected by the commissioners of both Canadian and American reservations. The cost of the trip is fifty cents.

THE CAVE OF THE WINDS.

THIS is one of the favorite attractions at Niagara. It is situated beneath the Central Falls, which are formed by the waters which flow between Luna and Goat Islands, and are at the extreme western side of the cliff.

It is formed out of the rocky cliff through the American cascade, the water, which has cut out the lower and softer formations, leaving the upper, a solid mass of Niagara limestone; this forms a roof, and one side of the Cave, the outside of it being a beautiful sheet of falling water, which here veils from view the outside world.

It is reached from Goat Island, where offices and reception rooms are found at the edge of the cliff, about one hundred dred feet west from the edge of the Central Falls. Here are provided dressing rooms for each sex, where a change of attire is made for the water-proof suits. Trusty guides and assistants are in attendance, one of the former having been on duty at the Cave for thirty-eight years, and has never known of but one accident in all that time; and this through the recklessness of the unfortunate man. Many think he deliberately committed suicide.

The Biddle stairs lead from the office to the pathway at the foot of the cliff; along this path the guide conducts the visitor to the edge of the Cave. Here the choice of two routes is given; one through the Cave and around the path which pass over little miniature falls, forming bridges and steps of waterways, and over rushing waters, and through a crack in the "Rock of Ages," which has been "cleft in twain," to the place of beginning. Or the route by the plank walk can be taken first and then the trip through the Cave of the Winds to the end of the journey.

We recommend the latter route, because if a feeling of timidity is felt about entering the Cave at first, it is apt to pass away by the time the Cave is reached from the other side, and the trip through it can then be taken with pleasure; and it will never be forgotten or regretted. It is the most awe-inspiring journey, for a short one, we have ever taken.

The Cave of the Winds is about one hundred feet long, twenty to seventy-five feet wide and one hundred feet high. The cataracts create a strong current of air, which in this particular locality finds the only accessible opening behind the Falls, and the air pressure in the Cave is so great that it is forced out with the force of a cyclone, the currents in the center have the velocity of a tornado, drifting the mist and spray from the Falls as a Dakota blizzard does the snow; hence the name, Cave of the Winds.

The price of admittance is one dollar, including attendance, guide and use of suits. The guide, however, usually anticipates the regulation tip.

(35)

EXCURSIONS BY WATER.

No Resort City on the continent offers more opportunities for pleasure trips than Niagara. When the tourist has completed the "sight-seeing" about the Falls, and starts out to view the historic country hereabouts, it is well to plan a trip by water to Toronto, the flower of the Province of Ontario. The traveler may make the journey by electric car from Clifton, on the Canadian side, to Queenston, along the brink of the Niagara Gorge, past Brock's monument and the battlefield of Queenston Heights, to the docks of the Niagara River Line; or may leave Niagara Falls or Suspension Bridge via the Gorge Route or New York Central observation trains to Lewiston, on the American side, taking the delightful ride down the Niagara river as it sweeps into Lake Ontario, seven miles away; or may continue the trolley trip on to Youngstown, on the American side, and embark from Niagara-on-the-Lake. If desired, the trip may be made via the Michigan Central from the Canadian side down to Niagara-on-the-Lake. The water trip is made at convenient hours by one of three handsome steel steamships of the Niagara Navigation Company, the Chicora, the Corona or the Chippewa, all of which are fitted out with complete electrical appliances and modern conveniences, and the trip across the lake, thirty-two miles from the mouth of Niagara river, occupies about three hours.

By leaving Niagara Falls early in the morning and embarking in the first steamer out from Lewiston or Queenston, the visitor has six hours in Toronto, and returns in time to connect via any of the several lines for the Falls, carrying memories of a delightful trip and a vigorous appetite for dinner.

(35)

Lewiston, from Queenston Heights.- Steamer "Chippewa."

Many excursionists who have not the time to visit Toronto embark at Queenston or Lewiston and make the river trip to the mouth of the river, where on either side of the bay are to be found the relics of the battles of the olden times, when the powers were struggling for control of this territory. On the American side is Youngstown, the scene of many bloody battles; and old Fort Niagara, once the seat of the French power in America, and which is now garrisoned. There is also to be seen a life-saving corps with exhibitions several days a week, and an excellent military band which gives daily concerts.

On the Canadian shore is Niagara-on-the-Lake, the Queen's Royal Hotel and resort, old Fort Mississauga on the Lake, old Fort George, erected by the British, and the new Chautauqua grounds and buildings. The return to the gorge may be made on any of the company's steamers, or by trolley cars of the Lewiston and Youngstown Frontier railroad, on the American side.

The steamship company has prepared a handsome little book, "How to See Toronto," giving details of the trips to be taken aboard their ships, and aiding in the selection of hotels while there, which will be sent those addressing Mr. John Foy, General Manager, at Toronto.

AN IDEAL SUMMER RESORT.

A MORE ideal spot could not be found for spending the summer months than that quaint little town, Niagara-on-the-Lake. Here every foot of ground is interesting to the lover of the early history of the struggles and triumphs of the British, and replete with relics and reminders of those bloody days. Opposite is the garrison U. S. Fort Niagara, sacred to the Americans, as a locality whose every nook and corner has a place in the history of those early battles for supremacy, so far back as the sixteenth century. Between old Fort Niagara and the Canadian shore the placid waters of the Niagara river just mingle with those of Lake Ontario. This spot first became important in 1750, when the British erected earthworks and attacked the French from here then known as Montreal Point, which movement resulted in the British gaining possession of Niagara after a siege of many days. Later in the century, the broad point jutting out into the green waters of the lake became an important military post, and was known variously as West Niagara, Butlersburg, Loyal Village and Newark; and here, after the organization of "Upper Canada," when the rebellion had settled the question of ownership of the United States, the first parliament was convened. Here, in 1756 the British erected Fort George, the earthworks of which are still undisturbed, and which occupied a prominent position among the British fortifications during the war of 1812. In the year named, the capital of Canada was removed from Newark to Toronto. During the war of 1812, Fort George was several times under heavy fire from the American fort and the village of Newark was destroyed entirely by orders of Gen. McClure of the United States army, when he retired from a temporary possession of the British fort in 1813. This wanton act resulted later in the almost total destruction of every building on the American frontier, when the British Indians were turned loose to avenge the destruction of the village. Further north, and on the very shores of the lake, was erected Fort Mississauga, merely a block house surrounded by earth fortifications, and which served as a support to Fort George in many engagements. Every foot of this ground was hotly contested many times, and relics of battles are often found in the immediate vicinity of the town now known by the romantic title, "Niagara-on-the-Lake." Now all is peace and harmony; both forts have been abandoned by the Canadians, and there only remain the ruins of the battlements to call to memory those days of strife. Here every year the quiet of the summer is disturbed by the annual encampments of the Canadian troops, who occupy the historical grounds that face Youngstown, and carry on mimic warfare for practice.

In the broad bay formed by the sweep of the river, the fleets of both American and Canadian yachts are gathered annually for regattas and pleasure excursions, and the waters are, during the summer season, continually dotted with white sails and pleasure boats. Back, beyond the shady streets of the little town, are many of Canada's best summer homes, and still further back, on the shores of the lake, are the buildings of the new Chautauqua.

Hotel Royal, Niagara-on-the-Lake.

QUEEN'S ROYAL HOTEL.

AMID this scene of repose and beauty is situated a charming summer hotel, "The Queen's Royal." The broad veranda of this hotel faces the green waters of the lake, and the clear waters of the Niagara where they meet, and on the other three sides are beautiful groves, flower gardens, shady walks, golfing links, tennis courts, and every provision which may add to the comfort of the guest in search of rest and pleasure. Just below is the sandy beach for bathing, and beyond are the best bass fishing grounds on the lake. On either side are the ruins of a fort, and but a few hundred feet distant is the landing of the ferries and the handsome steel steamers of the Niagara Navigation Co.

Messrs. McGaw & Winnett, of the Queen's Hotel, Toronto,

are proprietors of this resort, and are thoroughly up-to-date in caring for their guests. The hotel will comfortably accommodate 300 people, is lighted by electricity, has telephone and telegraphic connections, and every modern convenience. Rates $2.50 and upwards. Rooms or suites engaged in advance by addressing the proprietors.

Visitors to Niagara can reach the resort via the Gorge Route and the Lewiston-Youngstown electric road to Youngstown, and cross by ferry; the New York Central between the Falls and Lewiston where the steamers may be taken direct for Niagara-on-the-Lake; or if the Canadian trip is preferred, the steamers may be reached at Queenston via the Park & River Electric Road; or around the Queenston mountain over the Michigan Central. From Toronto the trip can be made by steamer or rail, as preferred.

(38)

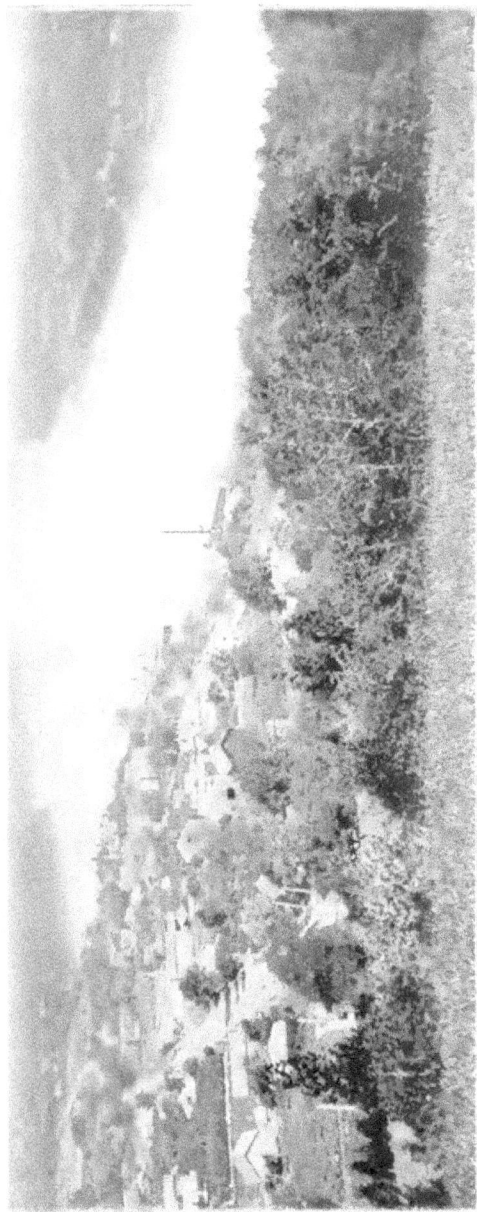

Queenston. From Brock's Monument, on the Heights.

Old Fort Mississauga, Niagara-on-the-Lake.

TRIPS OVER THE NIAGARA FALLS PARK & RIVER RAILWAY.

THIS popular electric line extends from a point on the upper Niagara river about a mile and one-half above Chippewa, where connection is made with the summer excursion steamers which make two round trips every day during the summer season between this point and Buffalo, to Queenston, a distance of about fifteen miles. Leaving the steamer's dock, the car passes the famous Chippewa battlefield, the little village of the same name, thence through the Dufferin Islands, into and through the Victoria Park, where the grandest views of the Falls are to be seen; passing the Table Rock elevator and bazaar where tickets are procured for a trip under the Falls, next the Dufferin Cafe and through the Park to Clifton, passing the popular hotel of that name, also the Hotel La Fayette, at the Canadian end of the new Suspension Bridge. It is at this latter point that most of our readers will take passage, either for a trip to Chippewa and return or in the opposite direction; tickets for the trip to Chippewa and return are 25 cents and are good for stop-over privilege at any point on the line and as often as it suits the pleasure of the passenger; and we advise all to take plenty of time, if they have it to spare, as the points of interest and the view are unsurpassed about Niagara.

The Horseshoe Falls from Above.

The trip to Queenston costs 50 cents for the round trip, or 75 cents for both round trips. In taking the trip to Queenston from this point, the line passes near the top of the cliff overlooking the gorge, passes the city of Niagara Falls, Ontario, the two handsome railroad bridges, to the incline railway leading to the Whirlpool Rapids, where from the long platform at the water's edge an excellent view of these troubled and excited billows is to be had; at this point the water rushes by with the rapidity of a fast train. The next point of interest is the Whirlpool, and from the car windows can be seen the upper and lower Whirlpool rapids and the Whirlpool, which is nearly circled by the double track of this scenic line; at the furthest point of the Whirlpool is a station,

and the best place to view these interesting scenes is only about one hundred feet distant. We advise a stop-over at this station, as the fifteen minutes between cars can be very pleasantly spent here.

The next point on the line is the Niagara Glen, or Foster's Flats, where the student of geology and lover of the wild and rustic in nature will find much to interest him; passing on with in sight of the gorge, the location of Bloody Run and the Devil's Hole can be seen on the opposite side of the river, and very soon after, Brock's Monument is reached; this English memorial was erected in honor of the brave general who fell at the foot of Queenston Heights, gallantly leading his men in a charge upon the Americans, who had previously carried the

(31)

The Upper Whirlpool Rapids, from the Battery Station.

heights, and dislodged some of the same soldiers he was urging forward to its re-capture. This they finally did, but not before General Brock had received his death wound. The view from this point is extremely beautiful; we have attempted to illustrate it with two views, but the camera to which we are indebted for the pictures we have reproduced, has failed to reach the distant points visible to the naked eye. The Duke of Argyle said that the view from Brock's Monument was worth crossing the Atlantic to behold.

From the heights, the road circles down the side of the mountain to Queenston, passing a tablet erected by the Prince of Wales to mark the spot where Brock fell, thence to the river's edge and upon the dock of the Niagara Navigation Co., where steamers can be taken for Lewiston, Niagara-on-the-Lake and Toronto, or the visitor can return to the Falls by the same route.

Ascending the Hill at Lewiston.

UNDER THE HORSESHOE FALLS.

FORTY years ago, the writer when a boy, took this trip, but at that time visitors were obliged to descend by an old stairway, now they go down the side of the cliff on an elevator with ease and safety, and instead of passing under the Falls, groping their way through blinding spray and mist as was the case at that time, a tunnel is provided for the passing beneath the Fall which reaches a platform directly behind the beautiful sheet of falling water.

The air pressure is quite strong, and the mist and spray are blown in dashes, very much to the surprise of the visitor, who is unable to account for the fierce blasts which drift beneath the Falls, this is and fast; but it is not so severe as to prevent the trip being an enjoyable one; and in the hot days of summer it is delightful, being cool, refreshing and agreeable as the shower bath, without its chilling effects.

No change of clothing is required, as the visitor has only to don a waterproof suit which fully protects the regular clothing. The fee of fifty cents which is charged for passing under the Horse Shoe Falls includes the use of this suit, the elevator and the services of a guide; tickets and suits are procured at the bazaar at Table Rock, nearly opposite the elevator.

(85)

The Three Sister Islands.

The City of Niagara Falls.—Birdseye View from Tower, looking East.

THE GEOLOGY OF NIAGARA.

To the geologist and student of this scientific study, the geology of the Niagara region, and especially that of the Gorge, opens up a book of nature full of interest.

In the Gorge, the geologist reads its age, and each stratum reveals pages of history, to them intensely interesting; even the school-boy who has just commenced to accumulate a knowledge of geology, readily concludes that the Falls have not always been where they now are, and he naturally wonders how long a time it will take them to reach Lake Erie, or the mouth of the river at Buffalo, if the present rate of recession is continued.

Professors Gilbert, Hall, Spencer and others, have all written and lectured upon this subject, and each has given valuable data and information.

The works of the former have been freely quoted from in the reports of the Commissioners of the Queen Victoria Niagara Falls Park, Ontario, and that of the latter by the Commissioners of the State Reservation, New York, (Prospect Park.) From both of these reports we are pleased to use extracts from the writings of these scientific gentlemen.

The geological points we think the reader most interested in are the age of the River, the Falls, and the Gorge, and the future recession of the Cataracts. Even on these points we have only briefly quoted those authorities, as our space will not admit of our going more deeply into this pleasing theme.

Prof. Gilbert writes as follows:—

"The middle term of our time scale, the age of the gorge, has excited great interest, because the visible work of the river and the visible dimensions of the gorge seem to afford a means of measuring in years one of the periods of which geologic time is composed. To measure the age of the river is to determine the antiquity of the close of the ice age. The principal data for the measurement are as follows:

(1) The gorge now grows longer at the rate of four or five

feet a year, and its total length is six or seven miles. (2) At the whirlpool the rate of gorge making was relatively very fast, because only loose material had to be removed. Whether the old channel ended at the Whirlpool, or extended for some distance southward on the line of the river is a matter of doubt. (3) Part of the time the volume of the river was so much less that the rate of recession was more like that of the American Fall than that of the Horse Shoe. Some suggestions as to the comparative extent of slow work and fast work are to be obtained from the profile of the bottom of the gorge. While the volume of the river was large, we may suppose that it dug deeply, just as it now digs under the Horse-Shoe Fall; while the volume was small, we may suppose that a deep pool was not made.

"Before the modern rate of recession had been determined, there were many estimates of the age of the river; but their basis of fact was so slender that they were hardly more than guesses. The first estimate, with a better foundation, was made by Dr. Julius Pohlman, who took account of the measured rate of recession and the influence of the old channel at the Whirlpool; he thought the river not older than 3,500 years. Dr. J. W. Spencer, adding to these factors the variations in the river's volume, computes the river's age as 32,000 years. Mr. Warren Upham, basing the same facts before him, thinks 7,000 years a more reasonable estimate. And Mr. F. B. Taylor, while regarding the data altogether insufficient for the problem, is of opinion that Mr. Upham's estimate should be multiplied by a number consisting of tens rather than units, thus estimates founded on substantially the same facts range from thousands of years to hundreds of thousands of years. For myself, I am disposed to agree with Mr. Taylor, that no estimate yet made has great value, and the best result obtainable may perhaps be only a rough approximation."

(45)

The New Suspension Bridge.

And Prof. Spencer reports that

"All attempt to reduce geological time to terms of years are most difficult, but the Niagara River seemed to be an easy chronometer to read, and yet we see that some utterances even this year are vastly farther from the mark than those made fifty years ago — the clock had not kept mean time throughout its existence. After this attempt at regulating the chronometer, investigators will doubtless carry the determinations to greater accuracy; but for the present I can offer this geological compensation. The Niagara seems a stepping stone back to the ice age. What is the connection between the river and the Pleistocene phenomena? The Lake epoch is an after phase of the Glacial period, and Niagara came into existence long subsequent to the commencement of the Lakes. If we take the differential elevation of the deserted beaches, and treat them as absolute uplifts in the Niagara district, with the mean rate of rise in the earlier portion of the lake epoch as in the later, then the appearance of Warren water in the Erie basin was about 60 per cent longer ago than the age of Niagara river; or about 32,000 years ago. The earlier rate of deformation was not greater than that during the Niagara episode, as shown by the deformation of the beaches, but it may have been slower, so that from 50,000 to 60,000 years ago Warren water covered more or less of the Erie basin. Before the birth of Niagara river, by several thousand years, there was open water extending from the Erie basin far into the Ontario, and all the upper lakes were open water with a strait

(42)

at Nipissing, but the northeastern limits are not known, and although they do not affect the age of Niagara, yet they leave an open question as to the end of the ice age, in case of those who do not regard the advent of the lakes as its termination. From these considerations it would appear that the close of the ice age may be safely be placed at 50,000 years ago."

"As has already been noted, the Falls was in danger of being ended by the turning of the waters into the Mississippi, when the cut through the Johnson ridge was effected. With the present rate of calculated terrestrial uplift in the Niagara district, and the rate of recession of the Falls continued, or even doubled before the cataract shall have reached the Devonian escarpment at Buffalo, that limestone barrier shall have been raised so high as to turn the waters of the upper lakes into the Mississippi drainage, by way of Chicago. An elevation of 60 feet at the outlet of Lake Erie would bring the rocky floor of the channel as high as the Chicago divide, and an elevation of 70 feet would completely divert the drainage. This would require 5,000 or 6,000 years at the estimated rate of terrestrial elevation. It would be a repetition of the phenomena of the turning of the drainage of the upper lakes from the Ottawa valley into the Erie basin.

"The computation of the age of the Niagara river,—based upon the measured rate of recession during 48 years; upon the changing descent of the river from 200 to 420 feet, and back to 320 feet; and upon the variable discharge of water from that of the Erie basin only, during three-fourths of the life of the river, to afterwards that of all the upper lakes,—leads to the conclusion that the Niagara Falls are 31,000 years old, and the river of 32,000 years duration: also that the Huron drainage turned from the Ottawa river into Lake Erie less than 8,000 years ago. Lastly, if the rate of terrestrial deformation continues as it appears to have done, then in about 5,000 years the life of Niagara Falls will cease, by the turning of the waters into the Mississippi. These computations are confirmed by the rate

and amount of differential elevation recorded in the deserted beaches. It is further roughly estimated that the lake epoch commenced 50,000 or 60,000 years ago, and there was open water long before the birth of Niagara, in even the Ontario basin, and that under no circumstances could there have been any hydrostatic obstruction to the Ontario basin since before the birth of Niagara Falls."

NIAGARA AS AN AIR-COMPRESSOR.

THE real cause of the recession of Niagara Falls, or at least the most destructive agent, has been discovered at last; and it is air—air completessed by the power of Niagara until it is forced to seek release in explosion. The discovery is of recent date, and only became public when we were about ready to issue this work.

Mr. J. C. Level, the owner of the Prospect Park carriages, and one of the proprietors of the Tower Hotel, has the honor of being the discoverer. Mr. Level claims no scientific attainments, but being a close observer and of an inquiring mind, he naturally wondered what caused the frequent outbursts of spray which occur at intervals of five to twenty seconds; something hurls these thin sheets and sprays of water high above the crest of the Falls, and with such velocity and energy as to indicate a power behind them resembling in its effects explosions.

"What is it?" That was the question Mr. Level propounded to the two scientific gentlemen from whose writings we have quoted in relation to the age of Niagara and the recession of the Falls. "If not an explosion, please tell me what it is," he further inquired. Each of these gentlemen were taken to the nearest point from which the phenomena could plainly be seen, and it was here these questions were asked. Both showed a great interest in the subject, and one of them remarked: "Mr. Level, you have made a great discovery; the other gentleman was not so candid, but said he would investigate only the

(47)

RUSTIC BRIDGES IN STATE PARK.

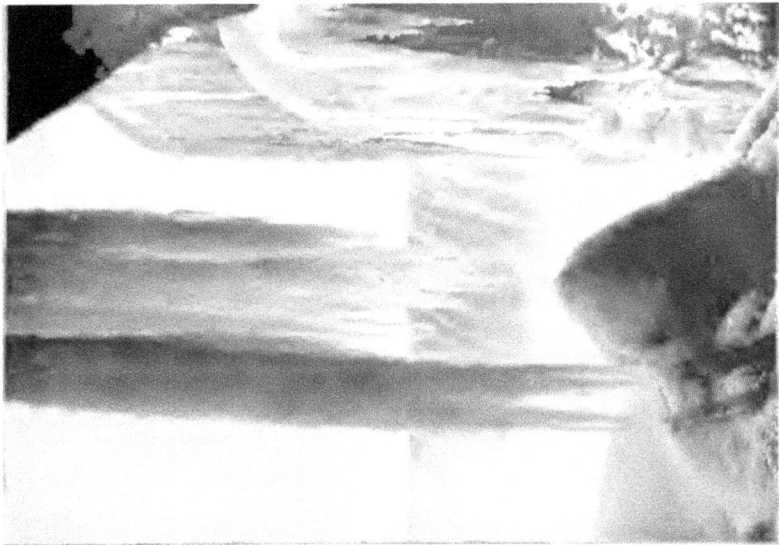

Winter Scene, Horseshoe Falls and Table Rock.

night previous, he had delivered a lecture on Niagara Falls, its recession, etc., in which he took about the same position as he had previously taken in his writings from which we have previously quoted.

Mr. Level's business required him to make frequent visits to Goat Island, and while in the vicinity of the Horseshoe Falls he has been in the habit of watching the air explosions and even listening for the sound of them at the base of the cliff on Goat Island, at a point the nearest to the Falls which can be approached with safety, where he thought he could detect the noise caused by the explosion. The results of his investigations have finally convinced him that his conclusions are correct.

We cannot comprehend why these men of science who have spent much time and thought on this subject did not make the discovery long ago; it seems so plain, the ocular evidence is so conclusive, that it should only require to have attention called to the facts in the case to convince any observer.

In the center of the Horseshoe Falls, where the larger body of water flows over the crest, and where the greatest recession of the cataract is observable, these explosions are frequent, and show by the height to which the spray is forced and the density of the water thus thrown high above the Falls (sometimes reaching a height of from 50 to 100 feet) that some great power is at work which is not visible, except in its effect.

Many know that there is effective power in compressed air; in many ways it is used to assist man; air drills have been in use for many years, and the force for their use transmitted long distances; as a motive power for street railways its application used a more recent date; but it has been successfully accomplished. Can this great power at Niagara be utilized? Who will be able to answer this question?

Is it compressed air that gives energy to this newly discovered power of Mr. Level?

This question is sure to be determined some time in the near future, because the widespread interest this discovery is sure to cause must bring about great research and investigation, and eventually, we trust, will result in its being utilized for the benefit of man.

Prospect Park in Winter.

(50)

Moonlight View in Prospect Park.

THE winter scenes at Niagara are extremely beautiful, and can hardly be surpassed outside of the Arctics. The mist and spray from the Falls, drift with the cold winds of winter and freeze upon every thing they come in contact with, the trees upon the islands, the parks, and upon houses adjacent to the cataract on each side of the river.

Our winter illustrations only faintly convey the real beauty of the views they represent, the sparkling gems of frost and ice are absent; these impart to the scenes a dazzling grandeur that cannot be represented. The great ice mountain, formed by the ice which flows down the river from Lake Erie, thrown over the Falls where it soon forms a gorge, rises often to a height of 100 feet. When it first accumulates it is clear and sparkling, but after a heavy fall of snow the open places are filled in and the rough edges smoothed over, giving the appearance of a mountain of snow rather than one of ice.

Long crystal icicles hang from different points, usually near the upper edges of the Falls, some reaching the ice below while others remain suspended without support except from above. These resemble in appearance the beautiful stalactites which are only found in large caves, but of course these icicles are more numerous and usually very much larger.

THE PARKS OF NIAGARA.

ON each side of the Niagara river are large tracts of land adjacent to the Falls, which were by act of the State of New York, on the American side, and that on the Canadian side by the Province of Ontario, condemned and then reserved for public parks. This is for the purpose of better protecting the great Cataract and affording the public at all times free access to the best points of observation from which Niagara Falls can be viewed. These legislative acts are the most important, in many particulars, of any laws passed for the benefit of this region.

They caused to be cleared away many unsightly things which have been replaced by those of beauty, and now the grand Cascades are surrounded by two beautiful parks. Many abuses have also been abated, and Niagara Falls now enjoys a better reputation as a pleasure resort than was possible under

American Falls, Winter View.

the old condition of affairs. It is true many individual citizens are now receiving fees from visitors for the privilege of gazing upon the beauties of Niagara from the best points; at which the Falls are to be seen, and which are now open, free to all. These citizens received a liberal compensation for their possessions; the amount paid for the lands by the State of New York being $1,433,429.50 for 107 acres. The Province of Ontario paid $49,683.24 for 154 acres included in their park; this was about $286,49.45 per acre, being about $10,000 less per acre than was paid for the lands embraced in Prospect Park on the American side by the State of New York.

These parks are maintained by the respective governments

on each side of the river, and are under the immediate charge of commissioners. The following named gentlemen are the commissioners for the New York State Reservation: Andrew H. Green, President; John M. Bowers, Robert L. Fryer, William Hamilton, George Raines, Henry E. Gregory, secretary and treasurer, Thomas V. Welch, superintendent. The latter is in personal charge, and has an office in the Park near Prospect Point. His management is very efficient and satisfactory. The commissioners of the Queen Victoria Park, Niagara Falls, Ontario, are: John W. Langmuir, chairman; Geo. H. Wilkes, Sen; F. Charlton, James Braunsfield, James Wilson, superintendent.

CARRIAGE SERVICE IN THE RESERVATION.

VISITORS to Niagara Falls who do not care to engage private conveyances may enjoy a delightful trip about the State Reservation in the Park carriages, which start every few minutes from the inclined railway building in the park and make the circuit of the reservation and the islands. The carriages seat comfortably a party of twelve persons. The drive is along the American rapids and through the park boulevards, past the rustic bridges, thence to Bath and Goat Islands. The drive about the island is one of the most attractive in all this great resort. The heavily wooded driveways lead first to Stedman's Bluff, whence stairways reach Luna Island and the Bridal Veil Falls; thence to the Cave of the Winds, Porter's Bluff and past Horseshoe Falls, along the Canadian Rapids to Three Sister Islands, to the Parting of the Waters, the Spring, and then back to the inclined railway building. Tickets, entitling the passenger to stop-over privileges at any point of interest, cost fifteen cents, the fee fixed by the commission controlling the park.

HISTORY, BRIEFLY TOLD.

THE history of the Niagara Frontier, so rich in historical events, can only be briefly told in the limited space at our command; but it is all so closely identified with Niagara Falls, it being the center of this field of historic interest, that we deem it essential to concisely relate an account of the most important battles and other occurrences of this region. We gladly refer the interested reader to the histories of "Old Fort Niagara" and "The Niagara Region," by Peter A. Porter, for a more extended account, as it is to this author, the historian of Niagara, we are indebted for most of the data from which our narrative is derived.

On the 6th day of December, 1678, LaSalle and his party landed at the mouth of the Niagara River, where, upon the point of land now occupied by Fort Niagara, he established a trading post. Soon after, he built quarters, which were protected by palisades, seven miles above at the head of navigation, on the beautiful location now occupied by the village of Lewiston.

Taking from his vessel, a brigantine of ten tons, a supply of tools, anchor, cordage and other materials to be used in the construction of a new boat, he had them carried overland to the mouth of Cayuga Creek, twelve miles distant,

Falls Street. Looking East from Monument.

and five miles above the Falls, over the route since known as the Portage Road.

Here he built the Griffon, the first craft larger than an Indian canoe that ever navigated the upper American Lakes.

The pretty little village of LaSalle, now occupies this site. The Griffon was completed in 1679, and set sail for the far west. Its outward passage was successful, but it was lost on the return voyage.

LaSalle was accompanied to this country by the French missionary, Louis Hennepin, to whom history is indebted for the first known picture of Niagara Falls, and also for one of the first general descriptions published, though it contained grave errors, the greatest of which was his estimated height of the Falls, placing them at five or six hundred feet.

For nearly one hundred years, the French and the Indians made most of the history of the Niagara Peninsula; the former holding the trading posts and forts, chiefly for the purpose of carrying on their fur trade. But they were never many years without their wars with each other, and the French frequently quarreling with the English concerning the possession of Niagara, which at that time meant that portion of this region lying between the mouth of the gorge and the mouth of the river. LaSalle's trading post at the latter point, which he converted into block houses, and named Fort Conti, was burned down within a year. A few years thereafter DeNorville built another fort, which though named after this French officer, was better known as Fort Niagara; and as such it still remains.

In 1759, the British forces under General Prideaux came to this region, determined to capture Fort Niagara; the army consisted of 2,050 men, including 750 Indians. They laid siege with great care and caution, building three lines of intrenchments for the better protection of their men during the engagement. During the battle General Prideaux was killed, and the command devolved upon Sir William Johnson, who continued the siege with vigor, and succeeded, after a bloody and fierce engagement which lasted several days, in

capturing the fort and its garrison. It was with the greatest difficulty that a general massacre of the French, by the Indian allies of the English, was prevented. Thus Fort Niagara became the stronghold of the British.

There were no engagements between the English and the American forces for the possession of Fort Niagara during the war of the Revolution, but by the treaty of 1783, the Great Lakes were recognized as the northern boundary of the United States; yet the English continued to hold possession of Fort Niagara and several posts on the southern borders of these lakes; but by the Jay treaty of 1794, they agreed to withdraw their garrisons from all these forts by June 1, 1796.

FORT SCHLOSSER AND THE OLD STONE CHIMNEY.

ONE and a half miles above the Falls stands an old stone chimney, on lands now owned by the Cataract Construction Company. It was built by the French in 1750, and was connected with a house which was at that time part of Fort Du Portage, afterwards known as Little Fort Niagara. These were destroyed by the French when they evacuated the place soon after the fall of Old Fort Niagara, and it was here that Fort Schlosser was built some time afterward. The houses with which the old stone chimney has been connected were destroyed by fire, but being substantially built the old chimney has continued to stand as a monumental history for 150 years.

Capt. Joseph Schlosser, a German by birth, but a British army officer, built Fort Schlosser in 1761. He remained here in command, and was later promoted to the rank of colonel; he died at the fort and was buried near by.

One of the houses built by the old stone chimney was used as a residence by Judge Porter in 1806-7 and 8. Later it was used as a tavern and accommodated visitors to the Falls; and was thus in use when destroyed by the British in 1813. The last house built by the old chimney was used as a farm house for many years.

Falls Street.— Looking West from Third Street.

THE BATTLE OF QUEENSTON HEIGHTS.

IN THE early days of October, 1812, Major General Van-Rensselaer collected an army of about 2,500 men at Lewiston, composed mostly of new recruits of New York militia. These became restless in camp and wanted to move on the enemy, and as it afterwards proved they were more anxious for a move than a fight.

On October 12, Colonels Chrystie and Fenwith, with 450 regulars, came up from Niagara, seven miles distant, and plans were immediately made for an attack on Queenston Heights early on the following morning.

Late on the evening of the day first mentioned, Colonel Winfield Scott, who had been in quarters at Black Rock, where the city of Buffalo is now situated, after a forced march with his regiment to Fort Schlosser, nine miles from Lewiston, hurriedly rode into the camp of General Vankensselaer and volunteered the services of himself and regiment. As all the plans had been previously arranged for the engagement, his offer was declined, but it was agreed that Scott should order his regiment forward from Schlosser to Lewiston and await results. They arrived at four o'clock the next morning, having spent a good part of the night on the march.

General Vankensselaer's plan was to cross the Niagara river before daylight, and thus secure a footing on the Canadian side before being discovered by the enemy. A scarcity of boats for transporting the men over the river, mismanagement and architects delayed these plans, and part of the land-ing was accomplished under a heavy fire from the British.

Colonel Scott not being able to secure boats in which to cross the river with his men, planted a battery on the American shore, and at daylight opened a vigorous fire upon the enemy.

The men who had succeeded in crossing the river were not much over one hundred in number, although they were afterwards re-enforced from time to time during the engagement. The few boats not disabled were inadequate for the demand, and the brave men in front of the enemy were under a very severe fire, but they formed in line and marched forward; in a few minutes every commissioned officer was either killed or wounded. They were soon joined by others, and among them Captain Wool, of the regular army, who being the highest in rank became the commander. Though himself wounded, he

Brock's Monument, Queenston Heights.

View from the Tower, looking South.

carried out the order of General Van Rensselaer, given as he was being carried from the field, "To mount the hill and storm the batteries." The hill was mounted and the batteries were carried. The enemy at this time was routed and retreated to a stone building under the hill near the water's edge.

Soon after, General Brock arrived, and later additional reinforcements for the British came from Fort George, at the mouth of the river. Brock soon collected the shattered forces, and made a vigorous attack upon the American position. In this engagement he was killed, but his army was totally successful, and completely routed the Americans; not, however, without meeting an enemy worthy of their steel.

Captain Wool ordered cut down an officer who had cowardly raised a white flag as a signal of surrender, and ordered shot another man who attempted to run, but who fortunately returned just as a musket was raised to shoot. Large numbers of the enemy were seen ascending the hillside, and soon after a hand-to-hand engagement ensued and several desperate attacks were repulsed; but finally many were driven over the steep bluff on the river side, only saving themselves from instant death by clinging to branches of trees and brush and lowering themselves to the water's edge. A few escaped, but many were taken prisoners. Among these captured was Colonel Scott, who during the fiercest part of the engagement had crossed the river to aid his countrymen, assumed command and fought bravely, an example, which if followed in sufficient numbers by the militia, who cowardly remained on the safe side while their comrades were fighting against a superior force within sight of their campfires, would have saved the day to the Americans and spared the disaster of defeat.

THE CAPTURE OF FORT GEORGE.

THE opening of the campaign of 1813 was very encouraging to the American forces. The combined attack on York (now Toronto) by the American army and navy under General Dearborn and Commodore Chauncey, was successful.

Arrangements were soon after made to attack Fort George, situated nearly opposite Fort Niagara, on the Canadian side, on the 27th day of May, by the same forces which had so brilliantly succeeded at York. General Dearborn was sick at the time and could not take command in person, but watched the movements of the land forces from the deck of one of the ves-

sels of the fleet, and Colonel Scott, who but a few months previous had been a prisoner at Fort George, was selected to lead. The guns from the fleet silenced the batteries of the fort, and the enemy, who had tried to repel the attack of the land forces, were beaten back at every point, and soon after completely routed, and in full retreat. Before vacating the fort, however, they blew up one of the magazines, which had been made nearly untenable by the heavy cannonading of the fleet. Scott followed the retreating army about five miles, and probably would have captured it had he not been recalled by superior officers. History says that the first order to discontinue the pursuit of the enemy was disregarded, Scott remarking to the young officer who delivered the order, "Your general does not know that I have the enemy almost within my grasp."

The official report of this engagement shows the loss of the Americans to be 17 killed and 45 wounded. The British loss was 90 killed and 160 wounded and 100 prisoners.

THE CAPTURE OF FORT ERIE.

ON THE 3rd day of July, 1814, quite early in the morning, Colonel Winfield Scott left camp at Buffalo in command of an expedition to capture Fort Erie. He had a brigade of infantry, supported by Hindman's artillery. They crossed the Niagara river to the Canadian side at a point below the fort, while Ripley's brigade landed above. Scott was in the lead, and accompanied by Colonel Camp, who volunteered his assistance, and they succeeded in reaching the shore before a gun was fired.

Coming up on both sides, the Americans attacked the fort with vigor and determination. After a sharp engagement the fort was surrendered, together with 170 men, including seven officers, who were taken prisoners and sent to the American side of the river.

THE BATTLE OF CHIPPEWA.

ON THE morning of July 4, 1814, Scott's brigade took up its line of march towards Chippewa. The Marquis of Tweedale and his forces contested every foot of the way, and a running fight was continued nearly all day; at dusk the British had been driven across Chippewa Creek. Both forces rested for the night, the British north of the Chippewa and the Americans south of Street's Creek.

General Brown, with additional troops, joined Scott's forces, and plans were discussed for an attack on the British at Chippewa. It seems that like arrangements had been made on the other side, and early on the morning of the 5th they commenced a sortie of light forces, and skirmishing continued all the morning, which was very hot and dry, and until the middle of the afternoon, with varied results.

At a time when the skirmishing had ceased and the fighting for the day seemed to be over, the British were seen advancing in full force; they had crossed the Chippewa into the Open field or plain which lies between the two streams, and it was here that one of the bloodiest battles of this war was fought, when the Americans engaged are outnumbered.

The Americans, under the gallant Scott, advanced over the bridge crossing Street's Creek into the open plain above mentioned, to meet them, and seeing a splendid opportunity for a heroic charge, he turned to his officers and men and said: "A British officer has made a remark this recklessness to the Americans, to the effect that they were very good at long range but could not stand hot shot or cold steel; I now call upon you to give the lie to this slander. Charge!" The charge was made with a will, and being well supported by artillery, the same heroic spirit soon extended along the whole line.

In the open field the two armies fought bravely, but the impetuous and gallant charge of the Americans could not be checked, and the British were completely routed; Scott followed the retreating army until it had crossed the Chippewa and reinforced their entrenchments.

The British forces numbered 2100 men, the American 1900 all told. The losses of the former were 138 killed, 319 wounded and 46 missing; total 503; American loss, 60 killed, 248 wounded and 19 missing; total loss 327. Grand total loss 830.

THE BATTLE OF LUNDY'S LANE.

SOON after the battle of Chippewa the British army retired to Burlington Heights, near the head of Lake Ontario. On the 10th day of July, 1814, the Americans moved their camp to Queenston and planned to capture the forts at the mouth of the river. For this they needed heavier cannon, and sent to Sackett's Harbor for them; but on account of the sickness of Commodore Chauncey they were not forwarded, and the attack upon the forts was abandoned.

Gen. Brown determined to attack the strong position at Burlington Heights, but concluded that it could be done with less loss if by feigning a retreat to Chippewa the British army could be induced to divide their forces. For this purpose the Americans returned to Chippewa.

It was Gen. Brown's intention to attack the British army at Burlington Heights, if his intended retreat failed to draw them out, and this was planned for the 26th; but on the 25th Gen. Brown received what was deemed reliable information that the enemy had crossed the Niagara from Queenston to Lewiston, 1000 strong, for the purpose, as Gen. Brown supposed, of capturing supplies on the way from Buffalo to Niagara. He at once resolved to return with his forces to Queenston and threaten the forts at the mouth of the Niagara, hoping thereby to cause the British to withdraw from the American side.

Scott's command was ordered forward while the other forces were preparing for the march. When about two miles from camp and one mile above the Falls several British officers were seen, and, as it proved, they were the advance guard of the enemy, which was concealed from view by a skirt of woods. Scott's forces consisted of only 1300 men, and he had positive orders to march quickly upon the forts by way of Queenston. Thinking that the British before him were only half the number he had whipped near by on the 5th of July, and that the rest of their army had crossed the river at Lewiston, he did not hesitate long as to what was best to do.

After sending an officer to inform Gen. Brown of the situation, he pressed his men forward through the woods that concealed the position of the enemy from sight. Here he found the British in full force with nine cannons in position on Lundy's Lane; whether better to retreat or do battle was a question to be decided quickly, and as Scott was always ready for a fight he chose the latter, believing it less hazardous for the whole army than a cowardly retreat.

The American stand was such a bold one that Gen. Raitt at once concluded that the whole army of Gen. Brown was before him, while in fact Scott's forces were less than one-fifth the number of the enemy. The reported crossing of the British to Lewiston proved to be untrue, and instead of their being reduced in strength by such a division, their army had been reinforced by large numbers under Gen. Drummond, who after this engagement was well advanced came up with part of his troops not already engaged and assumed command. It was after this general that Drummondsville, the little village situated upon the old battle-field of Lundy's Lane, was named.

It was nearly sunset when the battle opened, and very soon darkness found the two armies fighting fiercely for the advantageous position held by the British. It was nearly dusk o'clock before Gen. Ripley arrived with reinforcements, and at once took position on Scott's right, but finding it an unfavorable one, a movement was made nearer to the enemy, and meeting with a heavy fire from the artillery Ripley soon saw the necessity of capturing the battery, and asked Col. Miller if he could do it. "I will try," was his reply, and leading his command in the direction from which the deadly fire came they made a bayonet charge and drove the enemy from their guns, killing with the bayonet many who bravely stood their ground, and driving the living British soldiers from the hill which was the key to their position.

Four times did the British attempt to recapture their battery, and every time they were repulsed and driven back. In this dreadful strife, which lasted until midnight, the bravest men and officers of both armies were engaged; the losses were nearly equal, that of the British was reported to be 878, and the Americans at 860. Generals Drummond and Riall were both wounded, and the latter taken prisoner; of the American generals, Brown and Scott were wounded.

After the fighting had ceased and Gen. Brown was being carried from the field, he ordered Gen. Ripley to take command, collect all the wounded, remove the artillery captured, and retire to the camps they had vacated at Chippewa.

This order was only partly carried out; on account of the scarcity of horses and other appliances the cannon were left until morning, before which time the British, learning of the movements of the Americans, returned to the battle field, took possession of their old position and recaptured the nine pieces of artillery. Upon the strength of this, Gen. Drummond reported the result of the battle as their victory, and the English history so records it.

THE BRITISH ATTACK ON FORT ERIE.

ON the third day of August, 1814, Gen. Drummond and his army appeared before Fort Erie with the avowed purpose of carrying it by storm, but after examination and deliberation he concluded not to be too rash, and commenced a regular

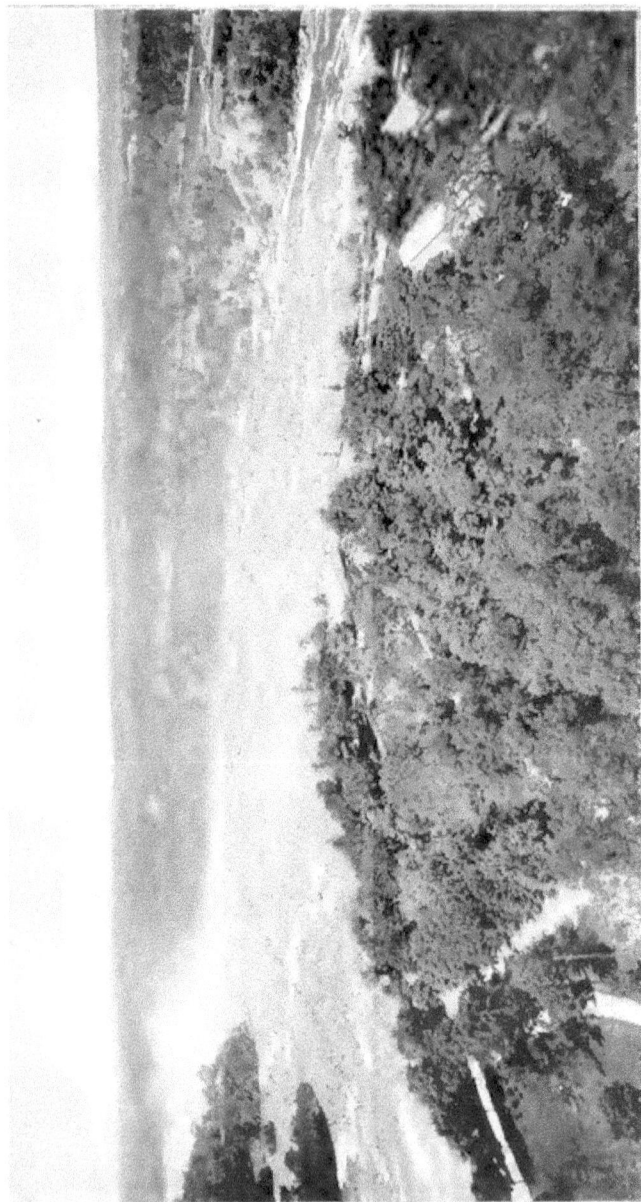

The Islands. Horseshoe Falls. Victoria Park, Canada. The Niagara River and Gorge.
Crest of the American Falls. Prospect Park

Birdseye View from the Tower, looking Southwest.

(62)

siege. From the 3rd up to the morning of the 13th they built earthworks and trenches, and attempts were made to cut off the supplies of the Americans. On the last day named they commenced a bombardment of the Fort, renewing it again on the 13th and continuing it until the evening without serious results; on the 15th they attempted to carry the Fort by storm, but were effectually repulsed with severe loss. Then another attempt was made on another side, with a like result; at the same time Gen. Drummond and the troops under his immediate command scaled the walls and got possession of part of the old Fort; soon after an explosion of cartridges occurred in one of the stone buildings, killing many and causing great confusion, during which the Americans drove the British out of the Fort. The official reports show their loss to be 57 killed, 313 wounded, and 539 missing; while the total loss of the Americans was only 84.

THE SORTIE FROM FORT ERIE.

AFTER the defeat of the British in their attack on Fort Erie, both armies remained quiet for some time. But on the 17th of September General Brown, who had recovered from his wounds received at Lundy's Lane, prepared to make a sortie upon the entrenchments of the enemy, hoping to raise the siege and drive the besiegers off. The fort had been invested for forty-five days. General Porter, who was familiar with the ground, suggested and planned this sortie, and was sent with his regiment to cautiously approach the position of the enemy. Quietly leading their way through the woods, to within short range of their works, and seeing that they were unobserved, he gave the order to charge, and in thirty minutes he succeeded

in capturing part of the works, two batteries and two block-houses; soon after another battery was abandoned and a magazine blown up; the cannon were spiked and dismantled and many prisoners taken. Thus the besieged became the besiegers, and the siege of Fort Erie ended. Very soon after this all fighting ceased upon the Niagara frontier, and the war of 1812 and 1814 was at an end.

Our space will not allow us to relate any particulars of the other engagements that occurred on the Niagara Frontier, and those already given are possibly more lengthy than they should be in a work of this kind; but they have been so related because of the fact that nearly all of the so-called guides published and sold at Niagara have uniformly given them from the British standpoint; our Americanism prompted us to examine into the history with the view of briefly relating facts as far as possible in our humble way.

Prospect Park, Early Spring.   Office of Supt. State Reservation.   Incline Railway Station.

Niagara Falls, Ont.

Niagara River and Gorge.

Milling District of the Niagara Falls Hydraulic and Manufacturing Co.

Birdseye View from the Tower, looking North.

CITY OF NIAGARA FALLS.

THE city embraces about eight and one-half square miles of territory, and has a population of from 17,000 to 20,000.

Four years ago it was known as the village of Niagara Falls, and Suspension Bridge, two miles distant, was a separate town, both together containing a population of 9,000. To-day they are both one, and the intervening space between the two former villages is now fairly well built over with business houses, the handsome dwellings of the well-to-do citizens, and the cozy cottages of the middle classes.

The present city limits extend far beyond the suburban settlements; and we have met those who believe that these outside tracts will never be needed for building purposes, and condemn the policy of speculators who control, by purchase or option, these districts for purely speculative purposes.

But this same trouble is to be met with elsewhere, and it only proves that somebody has confidence in "Greater Niagara." It is possible that these gentlemen have looked too far ahead, and in some cases have overreached, but that is purely a question to be decided by time. To the writer, who when a boy hunted quails upon the lands now embraced in the Central Park of New York City, it seems to be but a question of time when their expectations will be realized.

It is also claimed that property near the central part of the city is held too high, and that those looking for factory sites and homes are driven away in consequence. It would seem, however, that the policy of the two great power companies is too liberal to prevent the location of any desirable manufacturing enterprise or individuals seeking such an opportunity. These companies own large tracts of land adjacent and convenient to their plants, and within the city limits, which they purchased years ago at reasonable figures for just this purpose, and they are more than glad to make liberal terms as to factory sites or power.

The city formerly gained its notoriety from its beautiful Falls, and a larger number of visitors are attracted here to see them than go to any other resort in the United States, excepting perhaps a few summer attractions adjacent to some of the large cities, like Coney Island and similar places; which are visited in large numbers for a stay of a few hours only.

The Park reservations made on each side of the Niagara river in the immediate vicinity of the Falls, for the purpose of preserving their beauty and adding such conveniences as may be for the pleasure of citizens and visitors, must in the future enhance the popularity of Niagara Falls and increase its attractiveness as a pleasure resort.

But it is upon its growing reputation as a manufacturing center that its great future depends. The most sanguine predictions as to its growth fail to reach the point of possibility, as to its size and population, when the great power developments, now far past the stage of experiment, shall have completed the extensive improvements now under way, and in contemplation.

Our following remarks are based upon the accomplishment of these, i. e. the development of 550,000 horse power and its use at or near Niagara Falls.

We desire to be conservative, and have always to refrain from indulging in prophecy, but in this case we do not hesitate to express our private opinion, based upon the foregoing claims.

The development of this great power will in the future be more rapid than in the past. The Niagara Power Company have been five or six years in securing 15,000 horse power, but the work now under way to increase this to 50,000 will probably be finished within one year. By that time they may commence the second plant of the same size. This, of course, will depend upon the demand for power, of such demand, we are credibly informed, there is no doubt, as the 35,000 addi-

View, from Canadian Side, of the Milling District of the Niagara Falls Hydraulic and Manufacturing Co.

tional horse-power which will be ready for use before very long is nearly all spoken for in advance.

The Niagara Falls Hydraulic Power and Manufacturing Company have been still longer in their development, but they now have about 50,000 horse-power, and are pushing their work as fast as possible, and they anticipate doubling their capacity in a reasonable length of time.

When these 200,000 horse-power are in use, in the hundreds of factories that will be required to utilize it, what may then be expected of "Greater Niagara?"

This still leaves 250,000 horse-power to be anticipated later on. The world does not afford a parallel. Niagara Falls offers greater advantages as a manufacturing center than any other place in the universe.

Built upon their water power advantages, the great manufacturing centers, Holyoke, Lowell, Manchester, Lawrence, Lewiston, Me., and Minneapolis, had in 1890, a combined population of 388,502, and they have a total horse-power of 78,014; this gives them 4.98 persons for each horse power in use.

If this rate holds good here. Niagara Falls should have a population of 498,000 when the 100,000 horse power to be developed within the next year is utilized. When 350,000 horse power is developed this population would be 1,813,000. On this basis, when the total horse-power contemplated, 550,000 is furnishing power to manufacturing enterprises, the population of the greatest industrial center of the world, be it named Greater Niagara or Greater Buffalo, should be 2,739,000, and the city extend from Lake to Lake.

NIAGARA FALLS & SUSPENSION BRIDGE RAILWAY.

THE city of Niagara Falls is fortunate in possessing one of the best equipped and managed electric street car systems in the country, the seventeen miles of track being distributed throughout the portions of the city most in need of such transportation facilities, and touching at many points of interest, which makes it an enjoyable pleasure route for the visitor as well as the resident of this hustling city. There are two separate lines under one management, but a universal transfer is allowed at a five cent fare. The cars are of the Brill make, and the electrical equipment is that of the General Electric Company. The main lines start at Prospect Park, Falls street terminus, and a single fare carries the passenger out Falls street past the New York Central depot, past the great Hydraulic Tower Company's canal; the various mammoth manufacturing plants operated by electric power generated by the Niagara water; past the Niagara Falls Power Company's power houses; the site of historic Fort Schlosser, where the chimney erected by the French more than a century ago is still seven standing; past the landing place of the steamers from Buffalo; to Echota town, where a transfer takes the visitor through suburban Niagara and thence around to the old town of Suspension Bridge. Cars may be taken from this point to the Devil's Hole on the Niagara Gorge, or to the Buttery Whirlpool Rapids, from where the best possible view can be obtained of the great whirlpool rapids, or past the Rapids View, where an inclined railway heads to the head of the rapids below. This same line carries passengers to the Niagara University and the DeVeaux College. The main lines of the company pass the Convention Hall, it being but a brief ride from there to the principal depots or hotels.

The officers of the company are: I. T. Jones, president; Wm. B. Rankine, vice-president; C. B. Hill, secretary and treasurer, and J. C. Brewster, superintendent.

RAILROADS.

NIAGARA FALLS enjoys greater advantages in the way of transportation, for both passengers and freight, than any city of its population in the United States, as many of the most important trunk lines make it a terminus or have

branch lines to this city. The following railroads center here:

New York Central and Hudson River Railroad, Michigan Central Railroad, Rome, Watertown and Ogdensburg Railroad, Erie Railroad, Lehigh Valley Railway, West Shore Railroad, Grand Trunk Railway, Wabash Railway, Canadian Pacific Railway, Buffalo and Niagara Falls Railroad (electric), Niagara Falls and Lewiston Railroad (electric), Niagara Junction Belt Line, Niagara Falls Park and River Railway (Canadian side) (electric).

All these lines have access to the harbor and docks of the city over the tracks of the Belt Line, and are thus able to make connections with lake steamers drawing twelve feet of water or less. It is expected that the channel of the Niagara River will be dredged until there is a depth of water to the docks in this city of eighteen to twenty feet.

The tonnage handled by the railroads centering at Niagara Falls two years ago the latest figures obtainable was 10,338,430 tons. The total number of cars of freight handled was 706,067, including 27,140 cars of local freight.

The railroads of Niagara Falls have 92 regular daily passenger trains on their schedules for the summer of 1897, and in July and August there may be nearly as many more extra excursion trains; in the winter season they number about 80 daily trains. These figures give a very correct idea of the immense passenger business transacted by the roads centering there.

The report of the Commissioners of the State Reservation has a list compiled by Mr. T. V. Welch, the superintendent, of every excursion that arrived at Niagara the year previous to its publication; it shows the number of trains, cars and passengers, and the towns from which they came; we give herewith from this list a number of excursions for six months, May to October inclusive.

		No. Cars.	No. Visitors.
May	...	82	1,920
June	...	313	39,700
July	...	1,982	64,920
August	...	2,917	122,820
Sept	...	958	57,480
Oct	...	101	6,000
Total	...	4,615	276,000

Niagara Falls. N. Y., from the Observatory of the Hotel Lafayette, Canada.

Fig. 1

THE NIAGARA FALLS POWER COMPANY.

A Brief History of the Power Development at Niagara.

NOTE.—The data concerning the Niagara Falls Power Co. were compiled by Mr. L. A. Great at its Secretary, and that of the Niagara Falls Hydraulic & Manufacturing Co. by Mr. W. C. Johnson, its Chief Engineer.

TO souls sensitive to the beautiful and sublime, the plunging torrent of Niagara has appealed, by the stateliness of its stream, the brilliance of its boisterous rapids, and the deep glassy green of its silent, foreboding brink, as well as by its drop into the seemingly infinite depth, from which there comes to him who listens, the note of the welcoming abyss deeper than the diapason of any organ's pipe. To the weak and timid, there is danger and death in this resistless and remorseless tide, but to minds of dignity and self-restraint, the one sense to which the mighty cataract appeals most strongly is the "sense of power."

And why should it not be so? Nearly 6,000 cubic miles of water pouring down from the upper lakes, with 90,000 square miles of area, reach this gorge of the Niagara River at a point where its extreme width of one mile is by islands reduced to two channels of only 3,800 feet. Here, in less than half a mile of rapids, the Niagara River falls 55 feet, and then, with a depth of about 20 feet at the crest of the Horseshoe Falls, plunges 165 feet more into the lower river. The ordinary flow has been found to be about 275,000 cubic feet per second, and its force equal to the latent power of all the coal mined in the world each day (something more than 200,000 tons) representing theoretically 7,000,000 horse power, of which, according to Professor Unwin, the eminent English engineer and author, several hundred thousands of horse-power can be made available for practical use without appreciable diminution of the natural beauty of the Falls.

The idea of subjecting to industrial uses, some part of the enormous power of Niagara, has, since the early part of the eighteenth century, occupied the minds and stirred the inventive faculty of engineers, mechanics, and manufacturers. At an early day, the pioneers in the locality contemplated the probability, but were unable to demonstrate the practicability, of reducing this mighty force to obedient and useful service. They gave the name of Manchester to the early settlement, but the flourishing manufacturing center to be built by a utilization of Niagara's power remained but a dream.

In 1885, Thomas Evershed, an old, experienced engineer engaged in the service of the State, came to Niagara. After a conference with Mr. Evershed, several prominent citizens obtained a charter from the Legislature of New York, passed March 31, 1886, which by subsequent acts has been amended and enlarged. Mr. Evershed issued his first formal plan and estimate, which was described and discussed in "Appleton's Cyclopedia" for 1887, calling forth most adverse criticism and objections almost innumerable, which, in the light of subsequent successful achievements, have been fully answered. To convince capitalists that it would be commercially profitable to complete the development of Mr. Evershed's plans, required three years. It was demonstrated that the capacity of the proposed tunnel would be about 120,000 horse-power, exceeding the theoretical horse-power of Lawrence, Lowell, Holyoke, Turner's Falls, Man-

(69)

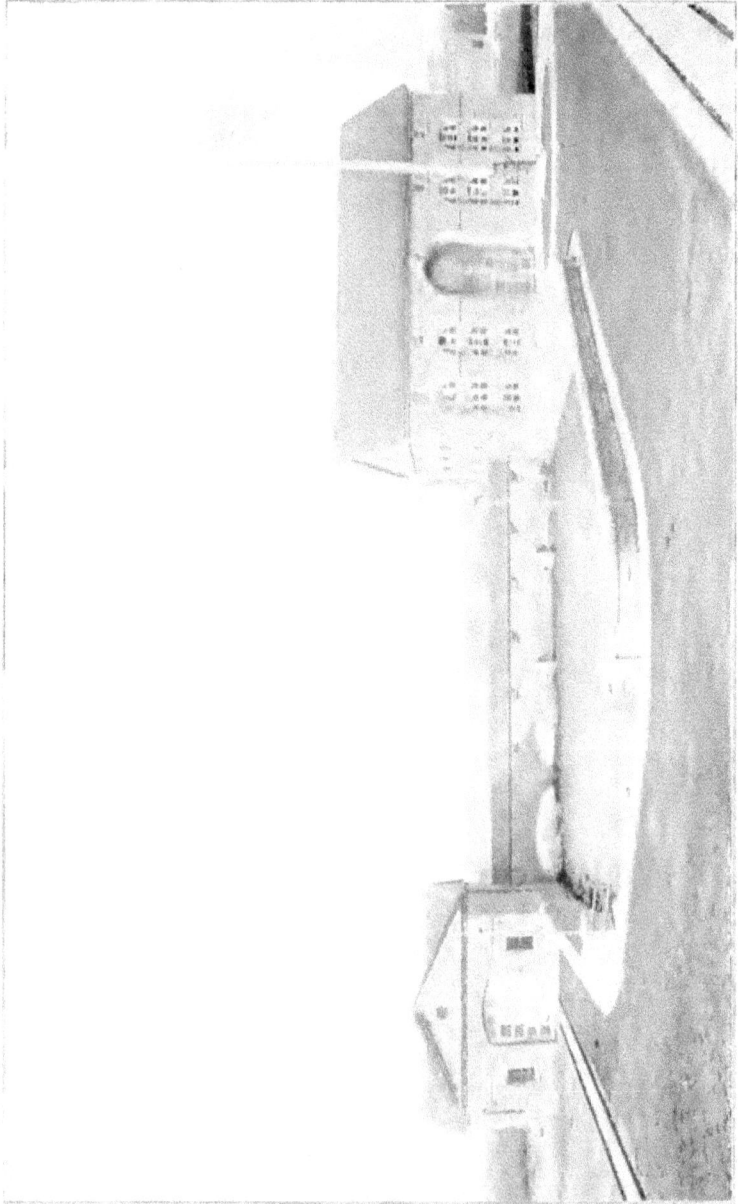

Transformer Building.   Supply Canal   Office and Power House.

Plant of the Niagara Falls Power Company.

chester, Windsor Locks, Bellows Falls and Cohoes; that it would largely exceed the actual developed power of all of these places, and Augusta, Paterson and Minneapolis in addition, representing more than a third of the power of all the water-wheels in use in the United States in 1880.

The advantages of Niagara Falls as a locality were fully shown, and the question whether water-power could be used in competition with steam was then discussed. After careful consideration the Niagara Falls Power company concluded that twenty-four hour steam horse-power is not produced anywhere in the world for less than $24 per annum, and that the cost of fuel represents but one-half the total cost. These considerations led to the organization of the Cataract Construction Company in 1889, which was the outgrowth of the zealous interest taken in the matter by the following gentlemen: William B. Rankine, Francis Lynde Stetson, Pierrepont Morgan, Hamilton Mck. Twombly, Edward A. Wickes, Morris K. Jesup, Darius Ogden Mills, Charles F. Clarke, Edward D. Adams, Charles Lanier, A. J. Forbes-Leith, Walter Howe, John Crosby Brown, Frederick W. Whitridge, William K. Vanderbilt, Geo. S. Bowdoin, Joseph Larocque, John Jacob Astor, all of New York City; and Charles A. Sweet, of Buffalo, most of whom have been officers of the Company. The plan finally determined upon comprised a surface canal, 250 feet in width at its mouth on the river a mile and a quarter above the Falls, extending inwardly 1,700 feet, with an average depth of 12 feet, serving water sufficient for the development of about 120,000 horse-power. The walls of this canal, which are of solid masonry, are pierced at intervals with inlets, guarded by gates. Some are used to deliver water to tenants putting in their own wheels and wheelpits, and ten inlets are arranged on one side of the canal to permit delivery of the water to the wheel-pit under the power-house, where dynamos placed at the top of the turbine shafts generate electricity for transmission to near and distant points. This wheel-pit is 178 feet in depth, and is connected with the main tunnel.

The tunnel has the purpose of a tail-race 7,000 feet in length, which serves a slope of six feet to the 1,000 feet. The tunnel has a maximum height of 21 feet, and a width of 18 feet, 10 inches, making a net section of 386 square feet. The slope is such that a chip thrown into the water at the wheel-pit will pass out of the portal in three and one-half minutes, showing the velocity of the water to be 26½ feet per second, or almost 20 miles per hour. In this great work 600,000 tons of material were removed; 16,000,000 bricks, 19,000,000 feet of lumber and timber were used, besides 60,000 cubic yard of stone; and 55,000 barrels of Giant American Portland cement, 12,000 barrels of natural cement, and 20,000 cubic yards of sand were used. Over one thousand men were engaged in the construction of this tunnel for more than three years.

The most careful consideration was given to the subject of the turbines, to be used, and also to the question of power transmission. In the winter of 1890, Mr. Adams, while in Europe, conceived the idea of obtaining information as to results obtained by engineers and manufacturers, not yet published, and in pursuance of this suggestion an International Niagara Commission was established in London in June, 1890, with power to offer $22,000 in prizes. The Commission consisted of Sir William Thomson (now Lord Kelvin) as chairman, with Dr. Coleman Sellers, of Philadelphia, Lieutenant-Colonel Theodore Turretini, of Geneva, Switzerland (the originator and engineer of the great water-power installation on the Rhone), and Professor E. Mascart, of the College of France, as members, and Professor W. C. Unwin, Dean of the Central Institute of the Guilds of the City of London, as Secretary. Inquiries concerning the best-known methods of development and transmission of power in England, France, Switzerland and Italy were made, and comparative plans were received from twenty carefully selected engineers, manufacturers of power in England and the Continent of Europe and America. All of the plans were submitted to the Commission at London, on or before January 1, 1891, and prizes were awarded to those consid-

Interior of Power House.

ered worthy by the Commission. The first important result was the selection of the designs of Faesch and Piccard, of Geneva, for turbines calculated to yield 5,000 horse-power each, and three of these wheels were built from these designs by the I. P. Morris Company, of Philadelphia, and are now in place. The question of turbines having been disposed of, the problem of transmission of power remained for solution.

After a careful study of the various methods of transmission by wire ropes, hydraulic pipes, compressed air and electricity, the Company in 1890, adopted the electrical system. The two-phase, alternating current dynamos employed were adopted under the advice of the Company's electrical engineer, Professor George Forbes, of London. In these the field magnet revolves instead of the armature, and three such dynamos of 5,000 horse-power each were made and installed by the Westinghouse Company, of Pittsburg.

During the summer of 1896, a transmission line 26 miles in length was constructed from Niagara Falls to Buffalo, and since November, 1896, the people of Buffalo have been enjoying the unique distinction of transportation in cars propelled by an unseen power generated more than twenty miles distant.

At the present time three 5,000 horse-power turbines and dynamos have been installed, but the rapidly increasing demand for power has necessitated the extension of the wheelpit and power house to more than three times their present capacity. The work upon this extension, sufficient to accommodate seven more 5,000 horse-power turbines and dynamos, has been in progress since June, 1896, and is being rapidly pushed to completion. Contracts for five additional 5,000 horse-power turbines and dynamos have already been awarded, and they will be installed as rapidly as the manufacturers can deliver them. With the full completion of the present extension. The Niagara Falls Power Company will have available 50,000 horse power, one-half of which is expected to be ready for delivery on or before December 31, 1897.

The Niagara Falls Paper Company, which was the first tenant of the Power Company, has been using 3,500 hydraulic horse-power for over two years. So thoroughly satisfied has this company been with the power furnished, that it has expended a million dollars in the erection of an additional plant and buildings, and is now using 7,200 hydraulic horse-power.

The Pittsburg Reduction Company manufacturers of aluminum, has for some time been using 3,000 electrical horse-power with perfect success, and of the other tenants of the Power Company using electrical horse-power, The Carborundum Company, manufacturing abrasives, and the Acetylene Light, Heat and Power Company, manufacturing carbide of calcium, have both recently evidenced their satisfaction with the electrical power furnished, by doubling the capacity of their plants.

The Niagara Falls Power Company has now contracts for present and future delivery of 7,200 horse-power of hydraulic power and 19,545 horse-power of electrical power. The cost of undeveloped hydraulic power is from $8 to $10 per horse-power, and upon the lands of the company, electrical power, two-phase, alternating current as it comes from the generator, is sold in large blocks at $20 per horse-power; a figure a trifle in excess being charged to purchasers of small blocks. These prices are for continuous 24-hour power.

Apart from the consideration of the superior reliability, cleanliness and convenience of electrical power over steam power, the above figures show conclusively the great economy to the consumer resulting from its use. It is doubtful if even under the most favorable conditions, steam power has ever been produced anywhere in the United States for less than $30 per horse-power for a 10-hour day; while the results of actual experience and tests show the average cost to be much greater. Recent tests made by a distinguished expert, of the cost of steam power at various plants located in different cities in the United States, show that the cost of power generated by steam, when produced under conditions most favorable

The Wheelpit Extension.

to economy, is $32.70 per horse-power for 11-hour power, while the average cost per horse-power is more than one-third greater. In a test of a large elevator at Buffalo, N. Y., he found that the cost of 3-hour power for 313 days per year was over $31 per horse-power. A test covering a year at one of the largest flouring mills in the State of New York—especially favored in location, and where every attention was given to secure economy in operating—the cost of 24-hour power produced by steam was found to be $45 per horse-power. The average cost of 24-hour power at the several different plants tested was found to be $63.60 per horse-power.

The 50,000 horse-power developed when the present wheelpit extension is completed, represents but one-half of the capacity of the present tunnel. A right of way for a second discharge tunnel has been secured, and when the demand for power shall render it necessary, the present plant will be duplicated. In addition to the 200,000 horse-power, for the development of which provisions have been made upon the American side, the Canadian Niagara Power Company—an allied corporation—now holds from the Canadian Government an exclusive franchise granting to it the right to develop upon the Canadian side, in the Queen Victoria Niagara Falls Park, at least 250,000 horse-power. Work upon the Canadian plant has already been begun, and by the terms of its franchise, the Canadian Niagara Power Company must have 10,000 horse-power ready for transmission and delivery on or before November 1, 1898. When fully developed, the American and Canadian plants will have a combined capacity of 450,000 horse-power.

When we consider that most factories use only from 5 to 50 horse-power, that Lowell, Mass., was built by 11,845 horse-power, that Minneapolis owes its marvelous growth and development to 25,000 horse-power, that the city of Buffalo uses less than 50,000 horse-power, and that the total power used in the State of New York does not exceed the combined capacity of these allied companies, then it is that the vast significance of the development of Niagara's power becomes apparent, and its meaning to the city of Niagara Falls and neighboring cities can be fully appreciated.

The first use of power at Niagara was about 1725, when the French erected a saw-mill, near the site of the Pittsburgh Reduction Company's upper Niagara works, for the purpose of supplying lumber for Fort Niagara.

In 1805 Augustus Porter built a saw-mill on the rapids. In 1807 Porter and Barton erected a grist mill on the river. In 1817 John Witmer built a saw-mill at Gill Creek. In 1822 Augustus Porter built a grist mill along the rapids above the Falls. From that time to 1885, when the lands along the river were taken for a State Park, a considerable amount of power was developed along the rapids by a canal which took the water out of the river near the head of the rapids and followed along nearly parallel with the bank of the river.

Mills were built between this canal and the river and a part of the fifty-foot fall between the head of the rapids and the brink of the falls was utilized. A paper mill was also built on Bath Island.

In 1847 Augustus Porter outlined a plan on which the present Hydraulic canal is built. In 1852 negotiations were commenced by Mr. Porter with Caleb J. Woodhull and Walter Bryant, and an agreement was finally reached with these gentlemen, by which they were to construct a canal, and receive a plot of land at the head of the canal having a frontage of 425 feet on the river; a right of way 100 feet wide for the canal along its entire length of 4,400 feet, which is through the most thickly populated part of the city; and about seventy-five acres of land near its terminus, having a frontage on the river below the falls of nearly a mile.

Ground was broken by them in 1853 and the work was carried on until 1858, when a canal thirty feet wide and six feet deep was finished. The location of the head of this canal was the best that could have been chosen. From the head of the rapids it is but a short distance to an island (Grass Island), which extends a considerable distance along the shore and for a considerable distance from the island the water is very shallow. In this short space, between the head of the rapids and the foot of Grass Island, the entrance of the canal was located.

Owing to the disturbed financial conditions occasioned by the War of the Rebellion, and other causes, it happened that no mills were built to use the water from the canal until 1870, when Charles B. Gaskill built a small grist mill on the site of the present flouring mill belonging to the Cataract Milling Company, of which Mr. Gaskill is President.

In 1877 the canal and all its appurtenances were purchased by Messrs. Jacob F. Schoellkopf and A. Cheshrough, of Buffalo, who organized the Niagara Falls Hydraulic Power and Manufacturing Company, of which Mr. Schoellkopf is still President.

Since that time the building of mills has gone steadily forward. The following is a list of mills using water from the canal:

WATER POWER.

	H. P.
Central Milling Co. (flour)	1,000
N. Wood Paper Co (paper and pulp)	500
Schoellkopf & Matthews (flour mill)	600
Pettebone-Cataract Manufacturing Co. (paper and pulp)	2,000
Cataract Milling Co. (flour)	100
Niagara Falls Waterworks	200
Thos. F. McGarigle (machine shop)	25
Cliff Paper Co. (paper and pulp)	2,500
Total	7,525

ELECTRIC POWER.

	H. P.
Pittsburg Reduction Co. (aluminum)	3,500
Niagara Falls and Lewiston R. R. Co.	400
Cliff Paper Co. (paper and pulp)	300
Lewiston and Youngstown R. R. Co.	300
Buffalo & Niagara Falls Electric Light & Power Co.	350
Niagara Falls Brewing Co.	150
Rockwell Manufacturing Co. (silver plating, etc.)	75
Sundry small customers in the city.	100
Francis Hook and Eye Co.	15
Kelly & McBean Aluminum Co.	15
Total	5,105

Cut No. 1.- Port Day (Mouth of Canal).

MECHANICAL POWER FURNISHED ON SHAFT.

	H. P.
Oneida Community, Ltd (silver plated ware and chains)..	300
Carter-Crum Co. (check book manufacturers)............	60
Total	360
Total Hydraulic Power sold	7,525
Total Electric Power sold	5,105
Total Mechanical Power sold........................	360
Grand total	12,990

Mr. Porter's contract with Woodhull & Bryant only conveyed the lands to the edge of the high bank of the Niagara river, and did not include the talus or slope between the edge of the high bank and the river, and only granted the right to excavate down the face of the bank one hundred feet.

At that time it was not considered that any higher head could ever be utilized, because it was not thought that wheels could be built to stand the pressure of a higher head, in fact none of the mills attempted to use more than fifty or sixty feet head. For this reason it happened that although the capacity of the canal as at first constructed was sufficient for some fifteen thousand horse-power, its capacity was exhausted and only about seven thousand horse-power produced.

The flouring mills of Schoellkopf & Mathews, Cataract Milling Company, Central Milling Company, the Pettebone-Cataract Paper Company, the City Water Works, and the factory of the Niagara Wood Paper Company leased the right to draw certain quantities of water from the canal and constructed their own wheel pits and put in their own water wheels.

Two different methods were adopted for constructing the pits for these various mills. In some cases a shaft was sunk in the rock at some little distance back from the edge of the bank, in which the wheels were placed, and a tunnel driven from the bottom of the shaft to the face of the bank for the discharge of the water after it had passed the wheels. In other cases a notch was cut into the face of the bank and the wheels placed in it.

In all cases turbine wheels of different makes, running on vertical axes were used.

In 1881 the Niagara Falls Hydraulic Power and Manufacturing Company put in a power plant for the purpose of supplying power to customers, delivered into their mills. The method adopted was as follows:

A shaft twenty by forty feet was sunk to a depth of about eighty feet, and about two hundred feet back from the face of the high bank from the bottom of this shaft a tunnel was driven to the face of the bank for a tail race. The water was conducted to the bottom of this shaft in iron tubes and used on different turbines running on vertical axes.

The power developed by these wheels (about fifteen hundred horse-power), was transmitted by shaft, belting or rope drive to various customers, all located within three hundred feet of the wheel pit.

In 1886 the Niagara Falls Hydraulic Power & Manufacturing Company secured a deed of portions of the slope between the high bank and the river, and have since secured other portions, so that they are now at liberty to use this slope for mills and power houses.

The advance in water wheel construction, and especially, the development of the possibility of transmitting power by electricity has made this one of the most valuable parts of their holdings.

In the spring of 1892 the Cliff Paper Company being desirous of increasing their plant, by adding a wood pulp mill, to use about twenty-five hundred horse-power, leased sufficient water from the Niagara Falls Hydraulic Power & Manufacturing Company, agreeing to take it from the tunnel through which the water was discharged from the outlet of the wheel pit just described.

For the purpose of getting the machinery requiring the largest power near to the wheels it was decided to build a mill on the lower bank near the water's edge, and to place the pulp making machinery in it, preparing the wood on the top of the bank, lowering it down ready for grinding and elevating the product.

To divert the stream of water flowing through the tunnel and confine it for use in the new mill, a short tunnel was driven into the face of the bank at a point about twenty feet below and twelve feet to the left of the mouth of the old tunnel.

From the mouth of the new tunnel, an iron pipe eight feet in diameter was laid along the slope of the bank connecting with the tube ten feet in diameter, in the basement of the lower mill. From this tube the water is brought to the wheels on the first floor. Provision is made for the discharge of water into the tunnel direct from the canal in case the discharge from the mills does not furnish a sufficient supply.

It was decided to use two wheels to develop the required twenty-five hundred horse power and to couple the shaft of the water wheel to the shafts carrying the stones used for grinding the wood.

It was therefore necessary that the wheels should be run at a speed of two hundred and twenty-five revolutions per minute. This requirement as well as the requirements of strength, precluded the use of any of stock wheels in the market and made a special design necessary.

Under the plans and specifications of W. C. Johnson, engineer for the Niagara Falls Hydraulic Power & Manufacturing Company, who was also engineer for the Cliff Paper Company, the wheels were built by James Leffel & Company, of Springfield, Ohio. (Interior of plant is shown in cut No. 2.)

In 1882 the Niagara Falls Hydraulic Power & Manufacturing Company commenced an enlargement and improvement of its canal. The plan adopted was to widen the original channel to seventy feet and to make the new part fourteen feet deep. The canal is cut entirely through rock below the water line.

The power for driving the drills on this work was obtained from an air compressor run by water power from the power station and transmitted along the line of the canal in pipes. The excavation was done by dredges, and the flow of water through the canal was not interfered with.

This improvement is now completed, and the canal has a capacity of about 3,000 cubic feet per second, giving a surplus power, after supplying the old leases, of about 10,000 horse power.

Work is still being carried on enlarging the canal to 14 feet deep and 100 feet wide. When this improvement is completed the canal will have a capacity of more than 100,000 horse power.

Since this improvement has been completed a new power-house has been built for the purpose of supplying power tenants.

For this new plant water is taken in an open canal from this hydraulic basin to a forebay thirty feet wide and twenty-two deep built near the edge of the high bank. From this forebay, penstock pipes built of flange steel eight feet in diameter, conduct the water down over the high bank two hundred and ten feet to the site of the power house on the sloping bank at the edge of the water in the river below the Falls.

The site of the power house was covered with broken and disintegrated rock, which had fallen from the bank during ages past, which covered the bed rock to a depth of from ten to seventy feet.

For the removal of this house material a Giant or Monitor, as it is termed was used. This is a machine throwing a stream of water from four to six inches in diameter, according to the size of the nozzle used, under pressure. It is very largely used in the western part of the United States for mining purposes, but has never been used in the east. This particular machine was purchased in San Francisco, Cal.

The water to supply this machine was taken from the canal, and the pressure of two hundred and ten feet head was sufficient to give a force which readily washed down all the loose material into the river, uncovering a bed of sandstone upon which the power house is built, and from which the material of which it is built was quarried.

The power house building will be one hundred and eighty feet long by one hundred feet wide and will contain 16 wheels, of about two-thousand horse-power each. One third of the length of the building is now constructed, and the second third is under construction.

The wheels in this power house work under a head of two hundred and ten feet, which is the highest head under which water has ever been used for power in the quantity used in this plant.

It was decided that water for the wheels should be supplied by a penstock leading from the forebay above described, vertically about one hundred and thirty-five feet to the top of the sloping bank, thence down the slope to the side of the station next to the bank, eight feet in diameter, connecting with a supply pipe ten feet in diameter, running horizontally along the center of the tailrace from which the wheels would draw their

Cut No. 2.—The Pulp Mill of the Cliff Paper Co.

water, by connections from the bottom of the wheel case to the top of the supply pipe. In this connection, which is five feet in diameter, valves are placed so that any wheel can be shut down independently of the others. The wheels standing directly over this trunk, discharge the water through draft tubes running down on either side of the supply pipe. (Shown in cut No. 3.)

Under general plans and specifications of the engineer, a contract was let to James Leffel & Company, of Springfield, Ohio, for supplying the wheels now in use. The description of the wheels is as follows:—

The wheel runners, in case of three wheels which run the generators of the Pittsburgh Reduction Company, and which run at a speed of two hundred and fifty revolutions per minute, are seventy-eight inches in diameter; in case of the other wheels, which run at three hundred revolutions per minute, 66 inches in diameter.

The rim of the runner (shown in cut No. 4) is the bucket ring, and is cast solid from gun metal bronze. On this rim are two sets of buckets taking water on face and discharging it at each side of the rim. The bucket ring is bolted to the spokes

Cut No. 4 — The Bucket Ring (Water-wheel).

by a supply pipe five feet in diameter. On the side of this case elbows are fitted to which the draft tubes are connected. The shaft passes out through these elbows through stuffing boxes. On the inside of the boxes lignum vitae steps are fastened, against which rings on the shaft work to prevent any motion in the shaft.

Each end of the water wheel shaft is rigidly coupled to a direct current generator, capable of developing five hundred and sixty kilowatts of electrical energy. The interior of the Power House as now built, is shown in cut No. 5.

The officers of the Niagara Falls Hydraulic Power and Manufacturing Company are:

Jacob F. Schoellkopf, President.
W. D. Olmstead, Vice-President.
Arthur Schoellkopf, Secretary and Treasurer.
W. C. Johnson, Engineer.

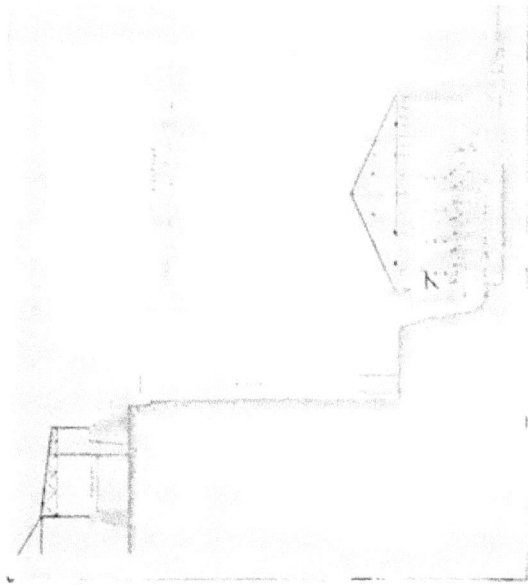

Cut No. 3. Penstock and Power House.

of cast iron center, the hub of which is keyed to the shaft of hammered iron twenty feet in length.

Surrounding the outside of the runner is a cylinder in which the gates are fitted. The gates are about twenty per cent less in number than the buckets. They are hung on steel pins and open by lifting one edge so that the direction in which the water enters the wheel is nearly tangential to the runner.

Each gate has two arms which are connected to the rings by means of which they are opened and closed.

This work is enclosed in a cylindrical case eleven feet in diameter and four feet long, which is connected to the penstock

Cut No. 5.—Interior Power House, Niagara Falls Hydraulic & Manufacturing Co.

Views on the Line of the Buffalo & Niagara Falls Electric Ry., near Cayuga Island, the Site of the Pan-American Exposition of 1900.

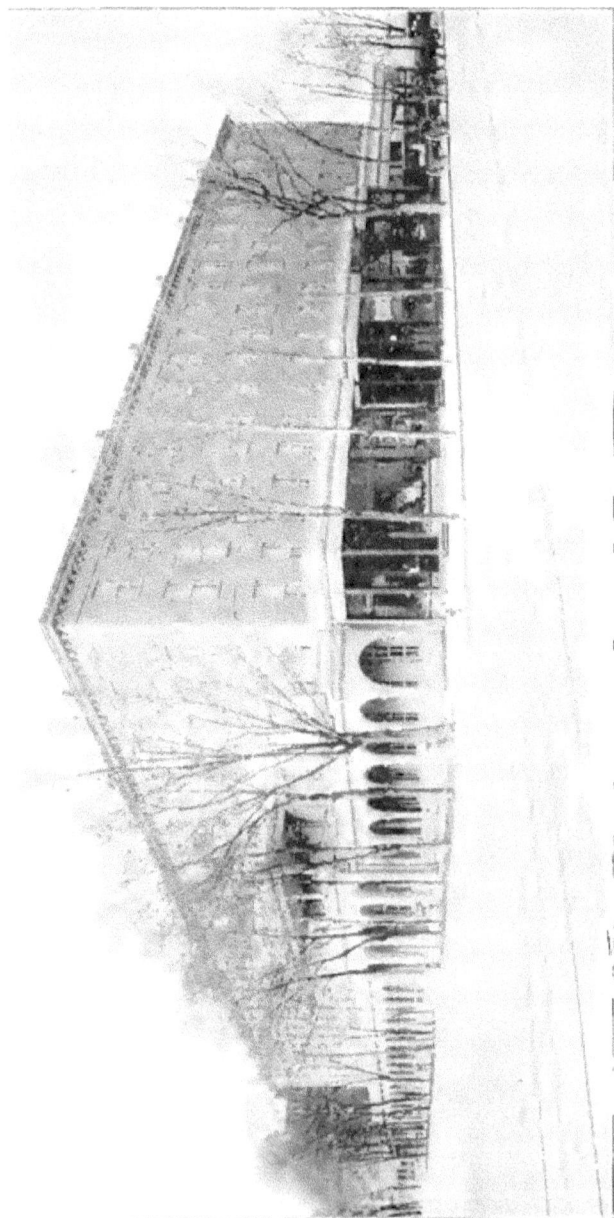

The International Hotel.

Table Rock in Winter.

THE INTERNATIONAL HOTEL.

THIS is the largest, best, and the leading hotel of Niagara Falls. It is situated on the block surrounded by Main, Falls and Bridge streets, and the New York State Park Reservation, the latter lying between the hotel and the American rapids. The river, rapids, and the beautiful islands are all in plain sight from the windows of the International facing the north, south, and west. The front of the building is on Falls and Main streets, the two principal thoroughfares of the city. The hotel has a nice little park of its own, into which the dining room extends, and upon which its windows open on two sides, making it delightfully cool and pleasant during the heated term. The house is substantially built of brick and Niagara limestone; the kitchen and laundry are in separate buildings, thus avoiding heat and odors, and insuring safety from fire.

The International can safely accommodate 600 guests, and during the months of July and August of each season, for nearly forty years, this popular hotel has been full. The season usually extends from May 15 to November 1, but the two hot months are the busiest ones for the hotels at the Falls. At this time the International is the center of attraction for the best class of visitors at this resort, and guests from other hotels gather here to meet their friends and enjoy the festivities of various kinds which follow each other in quick succession. It is a rule with the proprietor and managers of this hotel to amuse as well as entertain their guests, and therefore they have dances in the ball-room, musicals in the grand parlor, and private theatricals in their own park. An excellent orchestra is also engaged for the season, and several concerts are given daily.

Those contemplating visits during the busy months would do well to write for rooms and rates. A diagram will be furnished showing location of rooms, and rates will be made upon application. The regular rates are $3.50 to $5.00 per day, or $17.50 to $28.00 per week. All communications should be addressed to International Hotel Co., Niagara Falls, N. Y.

84

The Cataract House.

(85)

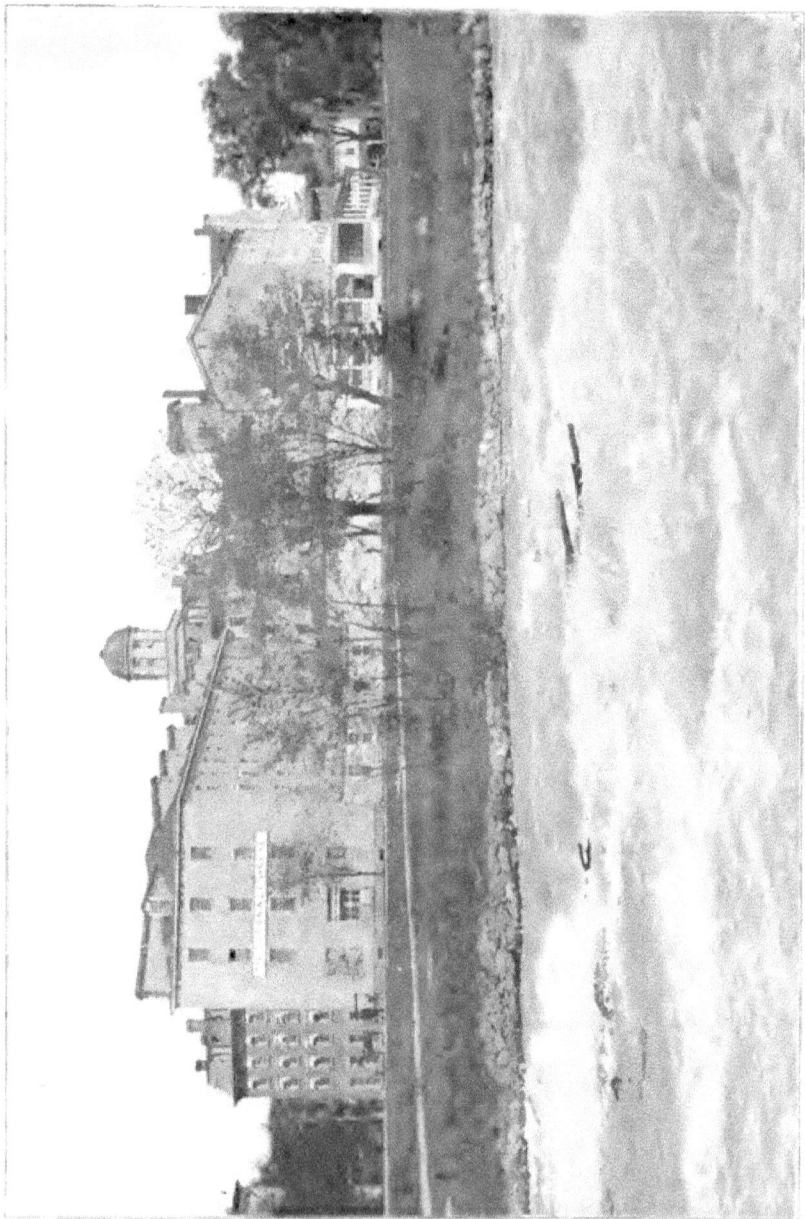

The Cataract House, from Bridge over the American Rapids.

(82)

"The Red Man's Fact"

THE CATARACT HOUSE.

THIS is one of the largest and oldest hotels in Niagara Falls. Part of the building was erected and opened as a hostelry in 1822. The Cataract, in all these years since its opening, has maintained a reputation for all that is excellent in a hotel, and has entertained a greater number of distinguished people, both foreigners and Americans, than most any other hotel in the United States.

The hotel is situated nearly on the banks of the Niagara, having only the narrow part of the Park of the New York State Reservation between. This Park also extends to the south side; the front being on Main street, the hotel and grounds occupying one whole block. Two wings extend towards the American rapids, and from the windows of all parts of the house facing south and west, magnificent views of the river and rapids are seen; also Goat Island and other adjacent landscape.

The Cataract House can accommodate between 400 and 500 guests, and, at times, in the busy season, every room is occupied by the best class of visitors that come to Niagara. In the map room and parlors of this hotel will be found oil paintings of the Falls, Rock of Ages, The Maid of the Mist (or white man's fancy), and the original of the legend of the "Indian Maiden and the White Canoe" (or red man's fact), together with maps of Niagara from its early history to the present time.

In a leaflet issued by this hotel we notice this: "The Cataract House neither seeks for nor caters to large excursions; but now, more than ever before, looks for, caters to and entertains that large class of persons who desire and will have the very best the market affords."

"The White Man's Fancy."

(87)

The Chilton House.

THE CLIFTON HOUSE.

No hotel in Niagara Falls is better known the world over than is the Clifton House, which occupies a prominent position at the very verge of the cliff facing both cataracts and at the entrance to the beautiful Queen Victoria Park.

Tourists, noblemen, and students from all parts of the globe have registered their names and enjoyed the hospitality at the Clifton, and during the season, this is one of the busiest resort hotels in America. The building is a substantial brick and stone structure and arranged with an idea for the comfort and convenience of its guests. The broad verandas extending along each story of the hotel furnish promenades and an unobstructed view of the gorgeous scenery of Niagara, and are veritable bowers of palms and flowers, lending additional attractions to the beauty of the picture of which the hotel is the central figure. In the hot summer days, the spray laden air as it kisses the clouds above, purifies and cools the sun-parched earth, and renders life at the Clifton most enjoyable.

The view from the Clifton is unsurpassed. The great gorge stretching away in the distance, the falls pouring streams of solid spray into the mad waters below, beyond the green landscape of Victoria Park with here and there a touch of red or yellow in the beds of flowers, the cool retreats where one can sit and watch the roaring waters fall, the whole making a picture that artist's brush has failed to reproduce in all its glory.

Close by is the Suspension Bridge, and a five minutes walk takes the tourist to Prospect Park, the American Falls, and several attractions which one must see in order to enjoy Niagara. Trolley cars pass the hotel and carry tourists to the Dufferin Islands, Chippewa, the Whirlpool, and Queenston.

The Clifton has accommodations for three hundred guests, and is under the personal management of Mr. G. M. Colburn, one of America's most successful resort hotel men. The cuisine of the Clifton is unexcelled and has made this house a popular stopping place. Mr. Colburn takes pride in the perfectness of arrangements for the enjoyment of his guests. The season opens in May and closes in November, and in this time, the register is positive evidence of the popularity of this hotel.

The Clifton is within easy reach of the Michigan Central, New York Central, and all other railways entering Niagara Falls.

On Goat Island.

TOWER HOTEL

PROSPECT HOUSE

THE IMPERIAL HOTEL

THE MALTERSHAM

THE TEMPERANCE HOUSE

THIS regular hotel is situated immediately opposite the New York Central Railroad Depot, on Second street, about two hundred feet north of Falls, the principal business street of Niagara.

The location is desirable on account of its nearness to the depot, business, banks, etc., making it especially handy for commercial men as well as visitors to the great cataract.

The hotel can accommodate over one hundred guests, and is one of the neatest and cleanest we have ever visited; a stay there will be found in every way satisfactory. Light and air abound, the house having no near building to cut out either; and in summer time an abundance of shade surrounds the building.

Mr. Hubbs, the proprietor of the Temperance House, is also the owner of this valuable property, and makes an agreeable host. His chief clerk, Mr. Wm. T. Phelps, has been with the house a number of years, and always makes friends among the guests of the house by his pleasing manners and genial disposition.

Mr. Hubbs owns two fine teams of horses and two carriages, one with two seats and one with three, which he keeps for the accommodation of the guests of the house who desire to ride

The Temperance House.

about the Falls and the best points of observation and interest, for which his charges are very reasonable; parties of two to five, one dollar per hour, or three dollars for the round trip. His drivers are very competent and trustworthy young men, and a ride with them will be found a very agreeable one.

Hotel rates one dollar and fifty cents to two dollars per day, or nine dollars per week.

The Columbia Hotel.

THE COLUMBIA HOTEL.

THE COLUMBIA is situated on the northeast corner of First and Niagara streets, one block north of Falls street, the principal business thoroughfare of the city, and about one block from the depots. It is a brick structure, four stories high, well arranged, nicely furnished, tidily kept and a very excellent hotel. The office, parlor and dining room are all on the first floor; the three upper floors are divided into sleeping apartments, en suite and single.

The situation of the Columbia, in the midst of private residences, quiet and retired, makes it a very desirable location for ladies and families who visit the Falls and prefer such surroundings, with their beautiful shady streets and walks.

The hotel has ample accommodations for about one hundred guests.

Mr. C. R. Phelps, the proprietor, for a number of years was connected with the Temperance House, six years as clerk and two as proprietor; and the friends he made there will be pleased to renew the acquaintance at the Columbia, where they will receive a hearty welcome. Mr. Phelps owns a handsome team of horses and a three-seated extension top victoria, and employs an obliging and trusty driver to take guests and private parties of two to five to the different points of interest about the Falls, over the new Suspension Bridge to Canada and the Victoria Park, from which the very best views of Niagara Falls can be had. His rates are one dollar per hour, or three dollars for the round trip, requiring about one-half day.

Hotel rates $1.50 to $2.50 per day, or $9 to $12 per week.

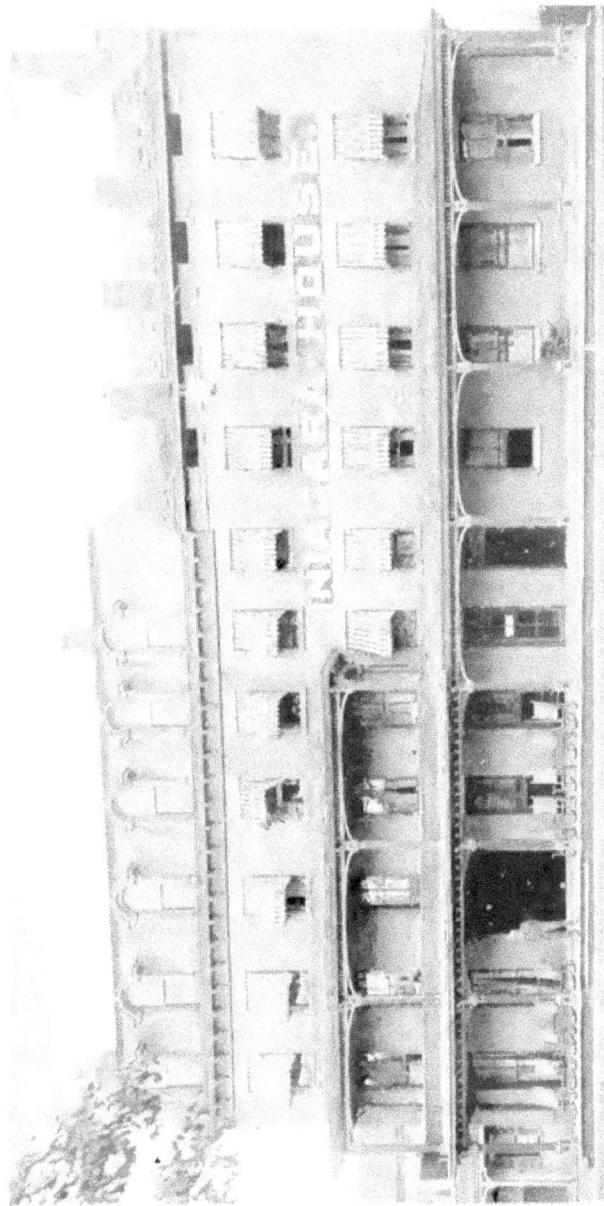

THE NIAGARA HOUSE.

THIS is one of the oldest hotels in Niagara Falls, having been established about the middle of the present century and being at one time the only competitor of the Cataract Hotel.

The rooms for guests number forty-two, and many of them open into suites of convenient number for families and tourist parties. The entire building has been refitted and refurnished throughout with the best of modern furnishings.

The location of the Niagara House is convenient to travelers, Prospect Park and the American Falls are but two minutes' walk from the Niagara House, and an excellent carriage service is maintained in connection with the hotel.

The Niagara House is located at 410-412 Main street, and is under the management of Mr. P. P. Pfohl, one of the popular young business men of Niagara Falls. He is assisted by Mr. Eugene Hall. Rooms may be secured in advance by addressing Mr. Pfohl.

SALT'S NEW HOTEL.

AMONG the best of the smaller hotels in Niagara Falls is the New Salt, 355, 357 Second avenue, between the New York Central and Erie depots, and which is under the management of Francis H. Salt and Stephen J. Tobey, two of the hustling young hotel men of the city. The house was erected by John Salt in 1899, Mr. Salt being succeeded by his son and Mr. Tobey one year later.

Salt's New Hotel has accommodations for seventy-five guests, and the conveniences are the best. The local electric cars pass the door and the cars of the Gorge railway may also be taken from this point.

Messrs. Salt & Tobey are particularly proud of the cuisine of their hostelry, and the house is deservedly popular. Rates $2 per day.

Salt's New Hotel.

THE PARK SIDE INN

IS LOCATED opposite the Queen Victoria Park and opposite the American Falls of Niagara. The Horseshoe Falls, a short distance above, are partially hidden from view by the beautiful shade trees of the Park, but in winter both falls are in plain view. It is one of the most charming spots on the Canadian shore, and in the season is a resort for wheelmen and tourists.

There are accommodations for fifty guests, and a private dining hall for regular guests. Rates $2 to $2.50 per day, and $10 per week.

The Park Side Inn is one block from the suspension bridge and convenient to depots. Lighted throughout by electricity.

Rooms secured in advance by addressing Mrs. W. J. McKoy, proprietor, or R. Laird, manager.

UNITED STATES HOTEL.

THIS hotel, established in 1870 at the southwest corner of Falls street and Second avenue, is one of the most fortunate in point of location. It is opposite the N. Y. Central depot, and one block from the Erie. Local electric lines pass the door, also cars of the Gorge and Niagara Falls & Buffalo Electric Railways. The hotel is three stories of brick construction, and accommodates 50 guests. August Rickert, manager, has had years of experience in catering to the public taste. Rates $2 per day.

THE HARVEY HOUSE

IS LOCATED on Third street, just north of the New York Central depot, and within one minute walk of the Erie depot. The Harvey is of brick construction, three stories high, and accommodates eighty guests. John Maloney is proprietor and owner. A free bus is maintained in connection with the hotel. Rates $2 per day. Rooms can be secured in advance for parties or families by addressing Mr. Maloney.

THE ELDORADO AT YOUNGSTOWN.

THE visitor to Niagara Fall who desires to pass a few hours or a few days at historic old Fort Niagara or the fishing grounds in the vicinity, should enjoy the hospitality of Landlord Steele of the ElDorado Hotel on the banks of the Niagara river as it sweeps majestically out into Lake Ontario. The ElDorado is a modern hotel, erected by Mr. F. C. Steele, formerly proprietor of the Brackett House at Rochester, and who knows, as the writer can attest, how to care for guests.

From the promenade over the dining hall the picture that surrounds the visitor is a remarkably beautiful one. Across the river is Niagara-on-the-Lake, seven miles above is Brock's monument looming up midst a blue sky and in a mass of

The ElDorado.

The Colonnade Hotel.

COLONNADE HOTEL.

OPPOSITE the Erie depot, and just one block from the Union depot, is situated the Colonnade Hotel, which was erected in 1886, and which, under the management of Cahavan & Donnelly, has become a popular hostelry.

The Colonnade is a three-story building, with accommodations for fifty guests, fitted out with electric bells and modern conveniences, and being but three blocks from the Falls of Niagara, is a convenient stopping place for tourists. Carriages may be secured from the management of the Colonnade. Rates $1.50 to $2 per day. Rooms engaged in advance by communicating with the management.

deep green verdure; the broad river winding in and out as it leaves Lewiston and carrying with it the sweet perfume of the flowers and woodlands. Below are the old fort and the life-saving station.

Above the hotel is a tower, free to all guests, from which on a clear day, one can view Toronto, 36 miles distant, and the vision sweeps the shores of the great lake and intensifies the pictures from below, many fold.

In front of the Eldorado pass the cars of the Lewiston & Youngstown Frontier Railroad, with connections every fifteen minutes for Niagara Falls via the Gorge Route or the N. Y. C. R. R.

Below is a boat house, where fishing tackle and boats are furnished for a day's sport, and nowhere can more delicious or gamey bass be found than here.

The hotel accommodates 100 guests. Rates, $2 per day and upwards, and $10 per week and upwards.

WILSON AND LAKE ISLAND PARK.

THIS is one of the most charming and delightful summer resorts near Niagara Falls; it is located on Lake Ontario, at the mouth of the Tuscarora river, ten miles from the Niagara river and sixteen miles from this city. The Tuscarora is navigable for some distance, and is protected at its mouth by two large piers running out into the lake forming a very pretty harbor for all kinds of vessels.

The islands formed by the turns and outlets of the river are beautifully wooded and dotted here and there by pretty cottages. Lake Island Park is owned by Walter N. Harris and Thomas C. Wilton, and is a very popular place, being visited throughout the summer season by people from all the surrounding country, Niagara Falls and Toronto; quite a number of the visitors to Niagara Falls spend one or two days of their sojourn in this region at this pleasant resort. The group of half tones which we use to illustrate Lake Island Park is from photos by our special artist.

The Tower Hotel, Wilson.

The tower hotel, a handsome half-tone of which may be found among our illustrations, is the best and most popular at Wilson; it is new and of modern construction, having all the latest conveniences. Mr. A. F. Bowker, the proprietor, is a most agreeable gentleman and has agreed to take good care of any of our friends who favor him with a visit.

Wilson is on the line of the Rome, Watertown & Ogdensburg Ry., and is also reached by a number of boats from points on the lake.

The Old Frontier at Lewiston.

NIAGARA IN ART.

NOTWITHSTANDING the fact that we have secured 130 photos, by Nevison, Curtis and Kurnz, from which the subjects for our engravings were selected and reproduced, we have also engaged a beautiful picture, in oil, of the Horse-shoe Falls, from Prof. F. Harold Hayward, the well-known artist of Mt. Clemens, Mich., who is now summering at Niagara, having

rooms at the popular Cataract House. Prof. Hayward, whom we have known favorably for several years, is an artist of note and ability, and we predict that his paintings of Niagara are sure to gain for him a national reputation.

We believe we have thus presented the finest and most completely illustrated publication ever devoted to the beauties of the Falls, although the most expert photographer admits his inability to more than vaguely portray the beauties of the region.

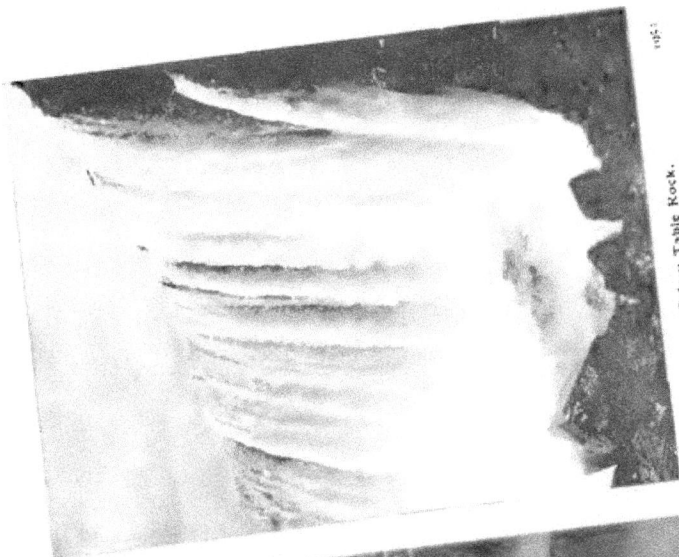

Below Table Rock.

Niagara Surface Coating Co., Manufacturing Paper for "Cutter's Guide."

EXPENSES AT NIAGARA FALLS.

SINCE the opening of the two beautiful parks, one on each side of Niagara, the Falls can be seen from every desirable point of view free of cost. The expenses generally are as cheap as at any resort.

Street cars to any part of the city, with transfers....$	.05
Toll, new Suspension Bridge......	.10
Toll, new Suspension Bridge, round trip......	.15
Falls to Steel Arch Bridge round trip......	.10
Park Carriages, Great Island, round trip......	.15
Incline Railway, round trip......	.10
Gorge Route to Lewiston......	.35
Gorge Route to Lewiston, round trip......	.60
Electric Line to Queenston, Canada......	.35
Electric Line to Queenston, Canada, round trip......	.60
Electric Line to Chippewa and return......	.25
Electric Line, round trip to both of the above......	.75
Burning Springs......	.50
Under Horse Shoe Falls from Table Rock, including guide and suit......	.50
Whirlpool Incline Railway, Canada......	.50
Battery Elevator (300 feet) to Whirlpool Rapids, each way......	.25 .25
To ascend stairs to top of Brock's Monument......	.50
Steamer "Maid of the Mist," round trip, including waterproof suit......	.50
"Cave of the Wind," including guide and waterproof suit......	1.00
Lewiston, Youngstown and Fort Niagara......	1.55
Lewiston, Youngstown and Fort Niagara, round trip......	1.00
Carriage hire, one to four passengers, per hour......	1.00
Carriage hire, one to four passengers, one-half day......	3.00
Carriage hire, one to four passengers, all day......	5.00
Three and four-seat conveyances, five to eight passengers, each per hour, 25c.; one-half day......	4.00
Regular meals at restaurants......	25 & .50
Regular meals at hotels......	50 to 1.25

HOTEL RATES.

American Side.

	PER DAY.	PER WEEK.
American Hotel	$1.00 to 1.50	$6.00 to 7.00
Walker House	1.00 to 2.00	6.00 to 8.00
Western Hotel	1.10 to 2.00	6.00 to 9.00
Central Hotel	1.00 to 2.00	6.00 to 7.00
Colonnade Hotel	1.50 to 2.00	7.00 to 10.00
Columbia Hotel	1.50 to 2.50	8.00 to 10.00
European Hotel	1.00 to 2.00	6.00 to 8.00
Exchange	1.00 to 2.00	6.00 to 8.00
Falls Hotel	1.00 to 2.00	7.00 to 10.00
Harvey House	2.00	7.00 to 10.00
Atlantique	1.00 to 2.00	7.00 to 9.00
Hotel Belgian	1.50 to 2.00	6.00 to 8.00
Imperial Hotel	2.00 to 3.00	8.00 to 15.00
Nassau Hotel	2.00	8.00 to 10.00
International	3.00 to 5.00	17.50 to 25.00
Hotel Schwartz	1.50 to 2.50	8.00 to 10.00
Niagara Falls House	1.50 to 2.00	7.00 to 10.00
Prospect House	3.00 to 5.50	
Solt's New Hotel	2.00	8.00 to 10.00
Temperance House	1.50 to 2.00	9.00
Tower Hotel	2.00 to 3.00	8.00 to 12.00
United States Hotel	2.00	7.00 to 10.00
Windsor House	1.00	5.00 to 6.00
Niagara House	2.00 to 3.00	9.00 to 12.00
Cataract House	4.00 to 5.00	17.50 to 25.00

Canadian Side.

	PER DAY.	PER WEEK.
The Hotel LaFayette	$2.00 to 3.00	$10.00 to 15.00
The Clifton House	3.00 to 4.00	18.00 to 25.00
The Windsor House	1.50 to 2.00	10.00
The Kosli	2.00 to 3.00	12.00
The American Hotel	1.50 to 2.00	7.00
The Arlington		4.00
Park Side Inn	1.00	10.00
Victoria Hall	2.00 to 2.50	6.00 to 8.00
The Imperal	1.00 to 2.00	4.00

WALTER W. STEELE, OSTEOPATHIC INFIRMARY,

Niagara Falls, N. Y.

THE DISCOVERY OF OSTEOPATHY.

THE revolt against our grandsire's medicine chest, with its ounce of quinine, its box of blue-mass, and its cup of ipecac, (not to mention its castor oil or its rhubarb) has been going on for years. But for a time no substitute could be found.

Health resorts, mineral waters, dietary and physical-culture tactics were all tried and found helpful but insufficient. The water cure gave promise of great good, but it too, was found wanting. In Faith Cure and Christian Science strength for a prostrate will was discovered, whereby this same will was enabled to resume control of bodily functions and often effect a cure.

But all of these systems were lamentably at fault, and every thoughtful physician felt that many diseases were beyond his control that should be within it. Still as a rule they bent their minds to the discovery of new drugs or more curious operations of surgery.

However one quaint character among them began, not a study directed by others along beaten paths, but an original study of bones. This physician was Andrew T. Still, M. D., and his first skeletons came from the historic Indian mounds of our great Mississippi valley. Death in his own family made him bitterly question the efficacy of materia medica; and caused him to turn his attention to the discovery of new and better curative methods.

Out of this search and research, not so much of books as of nature, grew piece by piece, fact by fact, the science of osteopathy. Parts of the science were used before Dr. Still himself understood the whole, or had discovered the fundamental principle; and when at last he felt that he had grasped the great idea of Life, Health and Cure, and held it in his hand ready to be applied, he had few adherents. Devotion to one idea had left him without money, friends or influence. But he gave the benefit of his idea freely to the poor, and it was not long till the rich were begging for it.

The labor soon became too great for one, and new operators were trained under the careful and critical eye of Dr. A. T. Still himself; and soon there grew up under his special supervision, a large school of Osteopathy. The demand for operators has been larger than the supply, and the labor thus thrown on the "Old Doctor," as he is lovingly called, has been a severe tax.

Whenever an operator is thoroughly fitted for his task, there is a place waiting for him, and patients eagerly gather about him, coming hundreds of miles.

WHAT IS OSTEOPATHY?

The name itself comes from two Greek words, Osteon, bone, and Pathos, feeling. The bones being the frame work, the foundation as it were, of the human being, and their proper adjustment being of prime importance, they were given the honor of naming the new science.

But the name is of small importance, the essence of the science being of principal interest. The foundation of this science is the idea expressed in the first chapter of Genesis, when God looked upon his handiwork and declared that it was "very good."

Man was made to be a well adjusted, unclogged machine, capable of action and labor until, worn by the friction of advancing years, work ceases and eternal rest begins.

8. When man is out of order, remove the obstruction and let nature do its work.

9. The efforts of nature at repairing are simply marvelous.

WHAT CAN OSTEOPATHY CURE?

The question is asked time and again, What can Osteopathy cure? As if they thought it a dose of quinine to be taken for chills and fever, or a Salvation Oil to be rubbed on for rheumatism!

But if Osteopathy is anything, it is the *Science of Restoring Health*. It has its fundamental principles and its truths built upon them. The application of these truths is the part of the D. O. Here it is that the science will grow and perfect itself, make new discoveries and perform new cures.

Its principles have already been successfully applied in the reducing of inflammation and fevers, in the quick relief of dislocations of hip, ankle and wrist joints; in relieving nervous diseases, eye and ear troubles; in unclogging and stimulating the digestive apparatus; in regulating the circulation; in quieting the nerves; and by means of a sensitive touch, locating the obstruction, pressure or obstruction, whatever or wherever it may be, that is causing the trouble.

So many stories are told of the wonderful working of this science in the curing of diseases, that the simplicity of its teaching arouses incredulity. But, as has been said, "The greatest Truths are the simplest; so are the greatest Men."

In the past six months Osteopathy has been recognized by the Legislatures of Vermont, Missouri, North Dakota, South Dakota, Michigan, Wisconsin, Colorado, North Carolina and Illinois. There are three schools in which the science is taught: The American School of Osteopathy, Andrew T. Still, Pres. Kirksville, Mo. Being the oldest and best known.

Walter W. Steele, D. O. the head of the Niagara Falls institution has performed many remarkable cures and references will be furnished upon application.

When this machine is out of order the Osteopath, by manipulation, replaces the slipped cog, wherever it may be; removes the pressure from muscles, nerves and ligaments; opens up the obstructed passages; by gentle measures, persuades nature to use the oil flasks of the body, and the human machine is in good order again and ready for its allotted task. Many a woman has made herself sick by the heavy exertion she has used to run a sewing machine when a readjusting of the parts, a little cleaning of the wheels, a tightening of the gear, or a drop of oil, would have done the work easily. No new fuel is added by the Osteopath to the fire already choked with ashes; nor the body mutilated when it should be simply straightened.

A few months ago it was the custom to laugh at the Osteopath's diagnosis of misplaced bones, but now that the X Ray is showing the slipped vertebrae of the spinal column incredulity must give way to belief, and it must be owned that the extraordinary cures are not mere accidents but the happy results of a known science.

You must understand that the Osteopath has no quarrel with surgery, recognizing that there are extreme cases where its use is a necessity. At many times, however, where surgery has heretofore been used, the trained fingers of the Osteopath do the work without its intervention; and humanity can scarcely be thankful enough for the science that rescues them from the misery of the drug and the knife.

The truths of Osteopathy as enumerated by its teachers are as follows:

1. Man is a machine.
2. He is created perfect.
3. When he gets out of order, the means of readjustment are within himself.
4. Drugs cannot create any part of this machine; nor replace any disturbed portion of it.
5. And this is all any physician is called upon to do.
6. He, by the use of his theories, cannot do it.
7. The Osteopath can and does.

LIMITED RATES TO NIAGARA FALLS.

City	Rate	City	Rate	City	Rate
Decatur, Ill.		Leavenworth, Kan.		Salt Lake City, Utah	
Denison, Tex.		Lewiston, Me.		Saratoga, N.Y.	
Denver, Colo.		Lexington, Ky.		San Antonio, Tex.	
Des Moines, Iowa		Lincoln, Neb.		Sandusky, Ohio	
Detroit, Mich.		Lima, Ohio		San Francisco, Cal.	
Deer Lodge, Mont.		Little Rock, Ark.		Sault Ste. Marie, Mich.	
Delaware, Ohio		Logansport, Ind.		Savannah, Ga.	
Duluth, Minn.		Cincinnate, La.		Seattle, Wash.	
Dubuque, Iowa		Mackinaw, Mich.		Selma, Ala.	
Dunkirk, N.Y.		Madison, Wis.		Shreveport, La.	
Elmira, N.Y.		Mansfield, Ohio		Sherman, Tex.	
Erie, Pa.		Marietta, Ohio		Sioux City, Iowa	
El Paso, Tex.		Mattoon, Ill.		Sioux Falls, S.D.	
Emporia, Kan.		Meadville, Pa.		Spokane, Wash.	
Evansville, Ind.		Memphis, Tenn.		Springfield, Ill.	
Fort Dodge, Iowa		Meridian, Miss.		St. Augustine, Fla.	
Fort Scott, Kan.		Mexico City, Mex.		Steubenville, Ohio	
Fort Worth, Tex.		Milwaukee, Wis.		St. Louis, Mo.	
Fort Wayne, Ind.		Minneapolis, Minn.		St. Paul, Minn.	
Galveston, Tex.		Mt. Clemens, Mich.		St. Joseph, Mo.	
Graham, W.Va.		Nashville, Tenn.		Superior City, Wis.	
Green Bay, Wis.		Nebraska City, Neb.		Syracuse, N.Y.	
Greenville, Tex.		New Haven, Conn.		Tacoma, Wash.	
Grand Island, Neb.		New Orleans, La.		Tampa, Fla.	
Grand Rapids, Mich.		Oil City, Pa.		Temple, Tex.	
Guthrie, Okla. Ter.		Olympia, Wash.		Texarkana, Ark.	
Harrisburg, Pa.		Omaha, Neb.		Terre Haute, Ind.	
Hartford, Ct.		Oshkosh, Wis.		Tiffin, Ohio	
Harpers Ferry, Va.		Palatka, Fla.		Titusville, Pa.	
Helena, Mont.		Palestine, Tex.		Toledo, Ohio	
Holly Springs, Miss.		Parkersburg, W.Va.		Topeka, Kan.	
Hot Springs, Ark.		Peoria, Ill.		Toronto, Canada (Boat)	
Houston, Tex.		Petoskey, Mich.		Toronto, Canada (Rail)	
Indianapolis, Ind.		Philadelphia, Pa.		Troy, N.Y.	
Iowa City, Iowa		Pittsburgh, Pa.		Urbana, Ohio	
Jackson, Mich.		Pittsfield, Mass.		Utica, N.Y.	
Jackson, Miss.		Piqua, Ohio		Vicksburg, Miss.	
Jacksonville, Fla.		Portland, Ore.		Vincennes, Ind.	
Junction City, Kan.		Portland, Me.		Vinita, Ind. Ter.	
Jamestown, N.Y.		Port Huron, Mich.		Virginia City, Nev.	
Kalamazoo, Mich.		Prairie du Chien, Wis.		Waco, Tex.	
Kansas City, Mo.		Pueblo, Colo.		Warren, Pa.	
Key West, Fla.		Red Cloud, Neb.		Washington, D.C.	
Knoxville, Tenn.		Richmond, Ind.		Waukesha, Wis.	
LaCrosse, Wis.		Rochester, N.Y.		Wheeling, W.Va.	
Lafayette, Ind.		Rock Island, Ill.		Williamsport, Pa.	
Lancaster, Pa.		Rome, N.Y.		Winnipeg, Manitoba	
Lansing, Mich.		Saginaw, Mich.		Xenia, Ohio	
Laredo, Tex.		Salem, Ore.		Zanesville, Ohio	

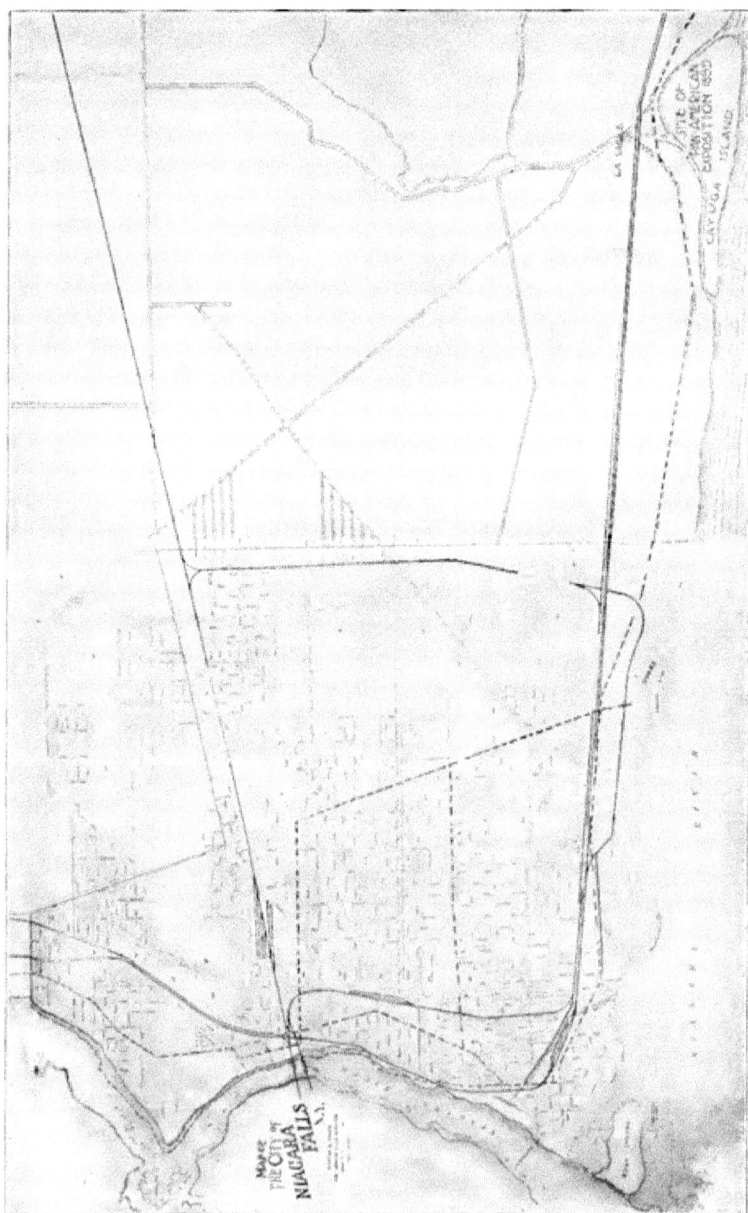

Map of
THE CITY OF
NIAGARA FALLS
N.Y.

LAKE ISLAND PARK, WILSON, N.Y.

NIAGARA NAVIGATION COS STEAMERS

MICHIGAN CENTRAL

THE NIAGARA FALLS ROUTE

A FIRST - CLASS LINE

FOR FIRST-CLASS TRAVEL.

THE ROUTE OF THE FAMOUS NORTH SHORE LIMITED.

CHICAGO, DETROIT.

NEW YORK. BOSTON.

AND TO THE

THOUSAND ISLANDS

AND

Rapids of the

St. Lawrence,

Canadian Resorts, Green Mountains, White Mountains, Portland, and the New England Coast, Northern Michigan Resorts.

MICHIGAN CENTRAL TRAIN AT FALLS-VIEW STATION

Send ten cents for a **SUMMER NOTE BOOK**, illustrated and descriptive

R. H. L'HOMMEDIEU, General Superintendent, DETROIT.

O. W. RUGGLES., Gen'l Pass'r and Tkt Ag't, CHICAGO.

SUMMER NOTE Book

MICHIGAN CENTRAL